The Future of Transdisciplinary Design

Lucienne Blessing · Ahmed Jawad Qureshi ·
Kilian Gericke
Editors

The Future of Transdisciplinary Design

Proceedings of the Workshop on "The Future of Transdisciplinary Design", University of Luxembourg 2013

Springer

Editors
Lucienne Blessing
Singapore University of Technology
and Design
Singapore, Singapore

Kilian Gericke
University of Rostock
Rostock, Germany

Ahmed Jawad Qureshi
University of Alberta
Edmonton, Canada

ISBN 978-3-319-06381-2 ISBN 978-3-319-06382-9 (eBook)
https://doi.org/10.1007/978-3-319-06382-9

This Springer imprint is published by the registered company Springer Nature Switzerland AG
The registered company address is: Gewerbestrasse 11, 6330 Cham, Switzerland

The Future of Transdisciplinary Design

Workshop Committee

Ahmed Jawad Qureshi
Kilian Gericke
Lucïenne Blessing

Scientific Committee

Alice Merner Agogino, UC Berkeley, USA
Guillaume Thomann, INP Grenoble, France
Jean-Yves Dantan, ParisTech Arts et Métiers, France
Eric Dubois, LIST, Luxembourg
Alex Duffy, University of Strathclyde, UK
Ashok Goel, Georgia Institute of Technology, USA
Tatjana Leblanc, Université de Montréal, Canada
Rainer Stark, Berlin Institute of Technology, Germany
Wim Zeiler, Eindhoven University of Technology, Netherlands

Preface

Companies are increasingly having to move boundaries in order to remain competitive. They need to address issues that challenge their core competencies or even challenge the state of the art on which their core business is built. They need to address issues outside their normal areas of competency, such as societal, ethical and environmental impacts of their business. All these issues have existed before, because the goods and services we develop do have an influence, on individual behaviour, environment and society, intended or unintended, short-term or long-term. What has changed is that (i). addressing these issues is socially and politically considered more urgent than ever before; (ii). consumers increasingly judge companies on the wider impact of their products, services and systems; (iii). Science and engineering has provided knowledge and technologies to address problems of increasing complexity.

Experience, both positive and negative, has shown that the grand challenges society is facing cannot be solved by technological means only. They require a broader perspective, involving scientific, technological, environmental, economic, political, societal, and ethical considerations and solutions. The focus in design is no longer solely on the product, service or system and their use value (performance) and prevention of environmental impact. Experience and interaction, effect on and role in society, global impact and transformation and a more holistic interpretation of value have become important. This requires involvement of a wide range of disciplines. However, many disciplines have not worked together, have not been involved in design and have over many decades developed their own cultures, education, experience, knowledge, motivation, priorities, values and view of other disciplines. Just creating teams of disciplinary specialists is likely to result in conflicts, misunderstandings, poor decisions, ineffectiveness and inefficiency in creating solutions. Collaboration requires mutual understanding and trust, crossing or even removing boundaries, as well as reflection, shared values and multi-perspective decision-making, despite disciplinary and cultural differences.

Many terms have been used to express collaboration between disciplines: interdisciplinarity, cross-disciplinarity, multi-disciplinarity, even pluridisciplinarity, all of which are considered beneficial. Do we need an additional term or is transdisciplinary just a buzz-word? Transdisciplinarity as a term is not new, but it took many years to become of interest. It is said to have been introduced by Piaget in 1969 who used it to refer to the *internal* dynamics of science, believing that "the maturation of general structures and fundamental patterns of thought across fields would lead to a general theory of systems or structures" (Klein 2004). Others refer to Jantsch who used the term in an OECD conference on education in 1970. Jantsch "envisioned a multi-level systemic coordination of research, innovation, and education". According to Klein (2009), Jantsch's focus on a common human and social purpose, i.e. an *external* purpose, for disciplines to work together to solve larger problems, has had the most influence and many have built on his work.

Early work on transdisciplinarity can be found in the area of philosophy and sociology of science, and considerable literature exists now on transdisciplinary research. The approach also features more prominently in sustainable development and educational sciences. Interest in transdisciplinary design is far more recent. Agreement on what transdisciplinarity is, how it differs from the other terms and how it should be applied in design, however, does not exist. Nor is there much disagreement, as so little has been written about transdisciplinary design.

This book brings together research into the transdisciplinary aspects of design, highlighting current issues and future challenges from the perspectives of processes, people and products. The aim is to introduce and establish research and development in the emerging field of transdisciplinary design in the international research community. The book presents ideas and mature research addressing the main issues pertaining to The Future of Transdisciplinary Design.

Leading researchers from 13 countries and different disciplines, including design, engineering, communication, information technology and psychology contributed to this book, following an International Workshop on The Future of Transdisciplinary Design (TFTD) held from June 24–25, 2013 in Luxembourg, and endorsed by the Design Society. A total of 44 abstracts were submitted (77 authors from 21 countries). After a double blind review by the members of an international scientific committee, 24 papers were selected and their authors invited to attend the workshop to share their views and discuss the central question: What is transdisciplinary design? The chapters were finalised and reviewed again after the workshop.

This preface does not include a definition or compilation of definitions of transdisciplinarity to allow the reader to experience the rich tapestry that is transdisciplinary design, as created by the authors. I hope this richness of perspectives will inspire the reader to explore transdisciplinarity in their own research, education and practice.

Singapore, Singapore
2020

Lucienne Blessing

References

Thompson Klein, J. Prospects for transdisciplinarity, Futures 36 (2004) 515–526
Thompson Klein, J. Unity of knowledge and transdisciplinarity: contexts of definition, theory and the new discourse of problem solving. In: Hirsch Hardon G. (ed) Unity of Knowledge in transdisciplinary research for sustainability – Vol I, 2009, EOLSS Publishers/UNESCO, pp 35–69

Contents

Transdisciplinary Design Approaches

Opening up Design Methodology

Kilian Gericke, Sebastian Adolphy, Ahmed Jawad Qureshi, Lucienne Blessing, and Rainer Stark

Abstract Contemporary product development has become increasingly transdisciplinary. Design problems often do not match the boundaries of a single discipline. The integrated use of tools, techniques, and methods, which are intended to support designers in their work, is a subject of design methodology. This paper presents a critique of the current state of design methodologies from a transdisciplinary perspective, and their industrial uptake. A case for an open community based approach for design methodology is made that considers contexts and mind-sets and provides a platform for consolidating design methodology in a dynamic perspective. The paper particularly addresses the following research questions: What needs should be addressed in the further development of design methodology? What developments are required to meet these needs?

Keywords Design methodology · Systematic design approach · Transdisciplinary · Product development

K. Gericke (✉)
Chair of Product Development, University of Rostock, Rostock, Germany
e-mail: kilian.gericke@uni-rostock.de

S. Adolphy
Humboldt-Universität, Berlin, Germany

A. J. Qureshi
Faculty of Engineering, University of Alberta, Edmonton, AB, Canada

L. Blessing
Singapore University of Technology and Design, Singapore, Singapore

R. Stark
Virtual Product Creation, Fraunhofer Institute for Production Systems and Design Technology, Berlin, Germany

L. Blessing et al. (eds.), *The Future of Transdisciplinary Design*,
https://doi.org/10.1007/978-3-319-06382-9_1

1 Introduction

Design problems often do not match the boundaries of a single discipline. Subsequently, design practice requires collaboration of designers from different disciplines. This collaboration from different disciplines can happen in different ways described as multi-, inter, or transdisciplinary. Currently these terms are used interchangeably with a degree of overlap. Transdisciplinarity, as opposed to multi-disciplinarity and inter-disciplinarity, concerns that which is simultaneously between disciplines, across different disciplines, and beyond all disciplines (Nicolescu 2005). Ertas et al. (2003) define trans-disciplinary design as the integrated use of the tools, techniques, and methods from various disciplines.

The integrated use of tools, techniques, and methods, which are intended to support designers in their work, is a subject of design methodology. Here, the term design methodology is used in order to refer to a specific approach to design, for example described in Pahl et al. (2007). A design methodology is "a concrete plan of action for the design of technical systems (…). It includes plans of action that link working steps and design phases according to content and organisation." (Pahl et al. 2007) The action plans are supported by methods.

Current design methodologies and methods are essentially mono-disciplinary (Gericke and Blessing 2011). Exceptions are a methodology for the design of mechatronic systems (Verein Deutscher Ingenieure 2004) and some initial support for the development of Product-Service-Systems (PSS) (McAloone and Andreasen 2004; Sakao and Lindahl 2009) in which the design of services (Shostack and Kingman-Brundage 1990; Bullinger et al. 2003) and products are to be integrated. The integration of different design disciplines in the notion of transdisciplinary design is not sufficiently considered in current design methodologies (Gericke and Blessing 2011), which is a limitation but might also guide their further development.

Design methodologies aim to structure the design process, to support planning of product development projects, and to provide support for related design activities, in order to avoid project failure. The use of a design methodology as a way of thinking is said to enhance the probability of a successful product development project (Hales and Gooch 2004). "*even though it has seldom been established 'scientifically' that design methods work, it is certainly sensible to use these tools sensibly, especially in situations in which the firm's own experience falls short, and the design process threatens to come to a standstill.*" (Roozenburg and Eekels 1995).

The discrepancy between current design practice and the current state of design methodologies have motivated the following research questions:

- What needs should be addressed in the further development of design methodology?
- What developments are required to meet these needs?

In this paper we provide an overview about critiques of current design methodologies, discuss on-going developments, and outline directions for the further development towards a transdisciplinary design methodology.

2 Current State and Critiques of Design Methodology

2.1 Consolidation is Required

> … classic Design Methodology has deficits in supporting current or even future development work that necessitate a substantial reformation. (Birkhofer 2011)

As reformation does not mean to start from scratch, it can build upon existing work. As much of the existing work is fragmented or related to a (discipline-) specific context, consolidation is required (Birkhofer 2006; Andreasen 2011).

A challenge for consolidation and for overcoming the boundaries of current mainly mono-disciplinary approaches is the lack of understanding of the different disciplinary design processes and of the differences and communalities of the various methodologies and methods to support these processes.

Analyses of design methodologies from different disciplines such as: Mechanical Engineering, Electrical Engineering, Software Design, Industrial Design, Service Design, Mechatronics, Product-Service-Systems, Building Design, and Systems Engineering show that the reviewed approaches show considerable communality across the domains (Gericke and Blessing 2011, 2012; Eisenbart et al. 2011). Most approaches propose a stage-based procedural process model. The activities composing the different stages are quite similar across the disciplines, even though the naming of the stages might differ (Gericke and Blessing 2012).

However, the disciplines also show important differences regarding: the prioritization of specific activities, the modelling approaches proposed to represent the different design states, the number of discipline specific design states and design activities, the terminology, the design methods and tools proposed (Gericke and Blessing 2011, 2012; Eisenbart et al. 2011, 2012; Eckert and Clarkson 2005). A "design state incorporates all the information about a design as it evolves." (Dym 1994).

2.2 Critique of Current Design Methodologies

A review (Gericke and Blessing 2011) of literature comparing and analysing design methodologies and process models indicates that systematic approaches are seen as suitable means in order to offer support for designers in different disciplines. Consensus exists that further development of current design methodologies and related process models, as well as further research are required, as certain aspects of designing are not sufficiently considered. No consensus exists on how detailed the methodologies should be and how detailed they can be. The review provides following critique on current design methodologies and process models:

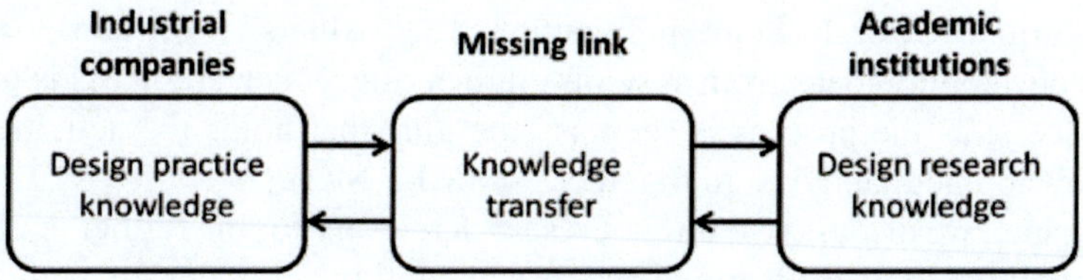

Fig. 1 Knowledge transfer—the missing link (Wallace 2011)

3.2 *Approaches to the Transfer of Design Methodology*

Up until the present day design methodology is largely transferred via lectures and textbooks and it seems that for many designers their academic training is the main source for knowledge about design methodology and design methods. Some attempts have been made to transfer the material in the textbooks into web-based method databases (TU Braunschweig; Lehrstuhl für Produktentwicklung TU München; RPK Universität Karlsruhe; Berkeley Institute of Design 2011; Design Council; TU Delft), providing additional features such as:

- Visualisation of entire methodologies;
- Searchability;
- Linking of methods to highlight similar methods or methods to apply sequentially;
- Templates for the execution of methods;
- Presentations for training purposes;
- Initial guidance for selection of methods.

Despite their benefits, the databases tend to have one or more of the following major limitations or areas for improvement:

- Discipline specific: the content only addresses only one or a few of the disciplines involved in design;
- Limited input from design practice: the database is created mainly by academics;
- Limited coverage: the content is usually provided by a small team, limiting coverage and depth;
- No maintenance: new methods and tools and user feedback are not included as many databases are not maintained;
- Repository only: the methods in the database are not linked to a methodology;
- Generic: the database does not support the transfer to a specific context.

4 Developments in Design Methodology

The need for consolidation, for overcoming the identified shortcomings of current design methodologies, and for addressing multi- and transdisciplinary design practices have stimulated new developments, most of which are still in a conceptual stage, requiring further research. Some propositions (e.g. Albers and Braun 2011) focus on further developing ideas from systems engineering. Much emphasis is put on the design process, as the process is seen as "the glue that holds the activities within product development and design together." (MMEP SIG).

In this paper, we discuss two lines of research relevant for the further development of design methodology. Both aim to enable a better transfer of design methodologies to design practice, but using a different paradigm. The first is the context dependent adaptation of design methodologies; the second is what we call the mindset approach.

The issues addressed in both lines of research are particularly relevant for a design methodology aiming to support transdisciplinary design practice.

4.1 Context Dependent Adaptation of Design Methodology

In this line of research design methodologies are interpreted as prescriptions for design work. The underlying assumptions are that prescriptive process models as proposed in design methodologies require a context dependent adaptation in order to serve as a basis for planning and design management, and that they should be followed as they represent best practice.

Miriam-Webster (Miriam-Websters) defines context as "the interrelated conditions in which something exists or occurs". Context factors are influencing factors, i.e. "'people or things having power', with power as 'the ability to affect outcomes" [Geis et al. 2008, p. 29] referring to (Lawrence and Lee 1984). In the context of design, these are the factors that influence the course of a design project.

The claim of design methodologies to provide support for a wide range of different contexts backfired. In order to cover a wide range of different contexts, the proposed process models, and thus the whole design approach, became rather abstract. The high level of abstraction resulted in the perception of being of limited use (Eckert and Clarkson 2005; Brooks 2010).

An approach suggested by different authors (Maffin 1998; Meißner et al. 2005) is to start with an abstract, context-independent approach and adapt it to a specific context. Lawson (1997) points out that the ability to manage this adaptation is one of the most important skills of designers.

Even though a context dependent adaptation is seen as a suitable means to substantially improve current design methodologies, only few contributions were made thus far, and it often remains somewhat fuzzy as to what context actually means. A challenge for empirical research on influencing factors is the vast amount of factors and their interdependencies which have to be considered when analysing the effects on product development (Gericke et al. 2013). The complex relationships between the influencing factors hamper a deduction of recommendations for a context-dependent adaptation of systematic approaches. As a consequence, adaptation is currently dependent on interpretation and understanding of the design methodology and the particular context by a particular designer.

4.2 *Mindset Approach*

The second line of research interprets design methodology slightly differently. Prescriptive models are considered not aiming to be a correct representation of how a product development process will proceed (Eckert and Stacey 2010). Eckert and Stacey (2010) argue that process models do "not need to be totally correct in order to be useful" because correctness does not guarantee usefulness. Thus, design methodologies and the provided prescriptive process models should be seen as guidelines—as heuristic methods, which need interpretation. Designers should not treat design methodologies in a dogmatic way, but rather opportunistically, and use them as guidance for their work (Bender and Blessing 2004). This we call a mindset approach.

Mindsets (Gollwitzer 1990; Gibson 1941) represent mental states of a person, leading to a preference of specific sets of mental processes depending on the particular mode of action "that produce a disposition or readiness to respond in a particular manner …" (Hamilton et al. 2011).

> A mindset is the proper understanding of a method's use in accordance with the designer's reality: interpretation of task, situation, execution, validation etc. and in accordance with the method's background and proper use. [Andreasen M.M., personal communication, 2013]

A challenge for the further development of design methodologies is to understand what enables the successful individual application of a design methodology in a specific context. This understanding is required in order to improve the training of design methodology and its transfer into practice and might also support the development of a support for a context dependent adaptation of design methodologies.

5 Opening up Design Methodology

Current design methodologies suffer from several limitations. In order to overcome these, the underlying concepts for the development and transfer of design methodology established in the past have to be changed. Aspects that have to be addressed are for example:

- **Disjoint research communities** focussing on their own engineering discipline, creating methodology islands, not sufficiently supporting transdisciplinary design practice.
- Transfer of the provided support via **static media** such as books and simple web-repositories.
- Waiting for **empirical proof of value** from applying design methodology.
- Lacking support for **adaptation**.

A concept for the further development of design methodologies addressing these challenges is to open up design methodology. The process of opening up design methodology is multidimensional and entails:

- Open for all disciplines (mechanical engineering, software engineering, product design, …), enabling the creation of a transdisciplinary design methodology.
- Open for practitioners and researchers, enabling a consolidation of existing support, best practices, new methods and tools, and research results.
- Open for active participation and feedback, enabling a dynamic evolution and continuous improvement.

5.1 From Disjoint Research Communities vs. Design Practice to an Open Design Community

As mentioned in the introduction, a large variety of design methodologies and process models exists (Gericke and Blessing 2012). Each is more or less specific to a certain discipline. This situation is not so much a result of the needs of design practice, but rather a consequence of historical development. Until today design methodologies are an output from fairly closed discipline-specific research communities. Transfer of knowledge from design practice to academia is currently done via empirical studies—by these same communities—of the design practices in their disciplines, and via practitioners entering academia, usually in their own discipline. The two communities, academia and practice, however, do not address the issues together.

To reduce the distance between design research and design practice communities, practitioners should be directly involved in the development of design methodology, i.e. an open design methodology should build on the knowledge of an open and active community including design researchers and design practitioners. Everyone interested has to be enabled to contribute actively in evaluation of the methodology and methods, their modification and extension.

As design practitioners have limited resources for extensive contributions to design methodology, mechanisms for low effort involvement have to be created, allowing practitioners to change and sculpt the open design methodology in a closed feedback loop. Such mechanism will not only provide practitioners opportunity to contribute to the building of the design methodology, but offer design researchers an invaluable test-bed for their research results.

Opening up design methodologies to academics and practitioners from a variety of disciplines is one step towards the creation of a transdisciplinary design methodology. Transdisciplinarity does not mean to strive for standardisation, but to identify and apply those practices, methods and tools that can be shared across the disciplines, and those that remain specific to a certain discipline.

5.2 From Static to Dynamic

Design methodologies tend to remain unchanged. The design methodology of Pahl and Beitz, for example, originates from the 1970s and only small changes were made over time. The situation in design practice, to the contrary, changed dramatically during these decades and will continue to do so. Hence a future design methodology has to be built with the capability to evolve with practice. Continuous involvement of practice will help to maintain applicability and usability of design methodology.

As continuous evolution of a methodology takes immense effort, the common approach of "create, evaluate, adapt" is not manageable by a small community of researchers. The integration of different disciplines and of design practitioners into an open design community could contribute.

An open design methodology should be created in an open web-based community enabling the required dynamics and a convenient stakeholder involvement. The internet offers excellent possibilities both for transfer of scientific knowledge into design practice and for feedback from design practice to design research.

5.3 From Empirical Proof of Value to Community Shared Benefits

The lacking acceptance and uptake of design methodologies has been related to the inability of the research community to demonstrate the (economic) value of the methodology they propose. The number and extent of empirical studies to be undertaken in order to unequivocally demonstrate the (quantitative) benefits of design methodology are not manageable. Demonstration of value through uptake in practice has not taken place either. Unfortunately, if there is no uptake there is no proof of value and if there is no proof of value there is no uptake.

Exploring alternative ways of demonstrating the value of proposed design methodologies might resolve this dilemma. In an open design community benefits as well as problems could be communicated between practitioners. Trust in recommendations of peers working in a similar context supported by empirical studies of academics might be more convincing than empirical studies alone.

5.4 From Generic to Context Dependent

Context dependent adaptation of a generic design methodology is required in order to meet the needs of design practice. However, relevant influencing factors and interdependencies are not known, and empirical deduction is practically impossible.

Hence, the required guidance for context-dependent adaptation has to be developed by different means. In an open community, the information on the adaptation of specific elements of the methodology will be built and shared as to minimize the effort for the individual and enrich the quantity and quality of specific adaptations. An open, web-based application with filtering and improvement of available support based on context-data and feedback could provide the additionally required guidance for the selection of suitable support.

6 Conclusion

Design methodology and its transfer from design research to design practice have to be improved significantly in order to meet the needs of the twenty-first century. Isolated, disjoint methodologies as evolved over the last decades are not sufficient for supporting transdisciplinary design practice, leading to rejection.

The challenges for the further development of design methodology, which arise from the evolvement of design practice, are manifold.

A consolidation of existing support and the further development of design methodology should be accompanied by an opening of design methodology to multiple disciplines resulting in a trans-disciplinary design methodology, and to researchers as well as practitioners, leading to a dynamic exchange of knowledge about new approaches as well as feedback from practice. Intensive research is still required of the various aspects involved.

References

Albers, A., Braun, A.: A generalized framework to compass and to support complex product engineering processes Albert Albers and Andreas Braun *. International Journal of Product Development. 15, 6–25 (2011).

Andreasen, M.M.: 45 Years with design methodology. Journal of Engineering Design. 22, 293–332 (2011).

Araujo, C.S., Benedetto, H., Campello, A.C., Segre, F.M., Wright, I.C.: The Utilization of Product Development Methods: A Survey of UK Industry. Journal of Engineering Design. 7, 265–277 (1996).

Bender, B., Blessing, L.: On The Superiority Of Opportunistic Design Strategies During Early Embodiment Design. In: Marjanovic, D. (ed.) 8th International Design Conference. Dubrovnik (2004).

Berkeley Institute of Design: The design exchange. (2011).

Birkhofer, H.: The Consolidation of Design Science—A critical review of the status and some proposals to improve it. In: Hosnedl, S. and Vanek, V. (eds.) Proceedings of AEDS workshop. Design Society (2006).

Birkhofer, H.: Summary—Holistic Ways to Supply, Extend or Replace Design Methodology. In: Birkhofer, H. (ed.) The Future of Design Methodology. pp. 211–217. Springer-Verlag London Limited, London (2011).

Birkhofer, H., Kloberdanz, H., Sauer, T., Berger, B.: Why methods don't work and how to get them to work. Proceedings of Third International Workshop—Engineering Design and Integrated Product Development. pp. 29–36 (2000).

Brooks, F.P.: The design of design: Essays from a computer scientist. Addison-Wesley, Upper Saddle River and NJ (2010).

Bullinger, H.-J., Fähnrich, K.-P., Meiren, T.: Service engineering—methodical development of new service products. International Journal of Production Economics. 85, 275–287 (2003).

TU Delft: Delft Design Guide. https://wikid.eu/index.php/Delft_Design_Guide.

Design Council: Design methods. https://www.designcouncil.org.uk/about-design/How-designers-work/Design-methods/.

Verein Deutscher Ingenieure: VDI 2206—Design methodology for mechatronic systems, (2004).

Dorst, K., Cross, N.: Creativity in the design process: co-evolution of problem-solution. Design Studies. 22, 425–437 (2001).

Dym, C.L.: Engineering design: A synthesis of views. Cambridge University Press, Cambridge (1994).

Eckert, C.M., Clarkson, P.J.: The reality of design. In: Clarkson, P.J. and Eckert, C.M. (eds.) Design Process Improvement A review of current practice. pp. 1–29. London (2005).

Eckert, C.M., Stacey, M.K.: What is a Process Model? Reflections on the Epistemology of Design Process Models. In: Heisig, P., Clarkson, P.J., and Vajna, S. (eds.) Modelling and Management of Engineering Processes. pp. 3–14. Springer, London (2010).

Eisenbart, B., Gericke, K., Blessing, L.: A framework for comparing design modelling approaches across disciplines. In: Culley, S.J., Hicks, B.J., McAloone, T.C., Howard, T.J., and Reich, Y. (eds.) Proceedings of the 18th International Conference on Engineering Design (ICED11), Vol. 2. pp. 344–355. The Design Society, Glasgow, Copenhagen (2011).

Eisenbart, B., Blessing, L., Gericke, K.: Functional Modellig Perspectives Across Disciplines: A Literature Review. In: Marjanovic, D., Storga, M., Pavkovic, N., and Bojcetic, N. (eds.) Proceedings of the 12th International Design Conference DESIGN 2012. pp. 847–858. Faculty of Mechanical Engineering and Naval Architecture, University of Zagreb; The Design Society, Glasgow, Cavtat (2012).

Ertas, A., Maxwell, T., Rainey, V.P., Tanik, M.M.: Transformation of higher education: the transdisciplinary approach in engineering. IEEE Transactions on Education. pp. 289–295 (2003).

Geis, C., Bierhals, R., Schuster, I., Badke-Schaub, P., Birkhofer, H.: Methods in practice—A study on requirements for development and transfer of design methods. In: Marjanovic, D., Storga, M., Pavkovic, N., and Bojcetic, N. (eds.) International Design Conference—Design 2008. pp. 369–376. Design Society, Glasgow (2008).

Gericke, K., Blessing, L.: Comparisons of design methodologies and process models across disciplines: A literature review. In: Culley, S.J., Hicks, B.J., McAloone, T.C., Howard, T.J., and Clarkson, P.J. (eds.) Proceedings of the 18th International Conference on Engineering Design (ICED11), Vol. 1. pp. 393–404. The Design Society, Glasgow, Copenhagen (2011).

Gericke, K., Blessing, L.: An analysis of design process models across disciplines. In: Marjanovic, D., Storga, M., Pavkovic, N., and Bojcetic, N. (eds.) Proceedings of the 12th International Design Conference DESIGN 2012. pp. 171–180. Faculty of Mechanical Engineering and Naval Architecture, University of Zagreb; The Design Society, Glasgow, Cavtat (2012).

Gericke, K., Meißner, M., Paetzold, K.: Understanding the Context of Product Development. Proceedings of the International Conference on Engineering Design, ICED'13. The Design Society, Glasgow, Seoul, Korea (2013).

Gibson, J.J.: A critical review of the concept of set in contemporary experimental psychology. Psychological Bulletin. 38, 781–817 (1941).

Gollwitzer, P.M.: Action Phases and Mind-Sets. In: Higgins, E.T. and Sorrentino, R.M. (eds.) Handbook of Motivation and Cognition—Foundations of Social Behaviour. pp. 53–92. The Guilford Press, London (1990).

Hales, C., Gooch, S.: Managing Engineering Design. Springer-Verlag, London (2004).

Hamilton, R., Vohs, K.D., Sellier, A.-L., Meyvis, T.: Being of two minds: Switching mindsets exhausts self-regulatory resources. Organizational Behavior and Human Decision Processes. 115, 13–24 (2011).

Howard, T.J., Culley, S.J., Dekoninck, E.: Describing the creative design process by the integration of engineering design and cognitive psychology literature. Design Studies. 29, 160–180 (2008).
Jänsch, J.: Akzeptanz und Anwendung von Konstruktionsmethoden im industriellen Einsatz—Analyse und Empfehlungen aus kognitionswissenschaftlicher Sicht, (2007).
RPK Universität Karlsruhe: Verbundprojekt. https://imihome.imi.uni-karlsruhe.de/map.html.
Lawrence, P.A., Lee, R.A.: Insight into management. Oxford University Press (1984).
Lawson, B.: How designers think: The design process demystified. Elsevier/Architectural Press, Amsterdam (1997).
Lehrstuhl für Produktentwicklung TU München: CiDaD—Competence in Design and Development. https://www.cidad.de.
Macmillan, S., Steele, J., Kirby, P., Spence, R., Austin, S.: Mapping the design process during the conceptual phase of building projects. Engineering, Construction and Architectural Management. 9, 1–7 (2002).
Maffin, D.: Engineering Design Models: context, theory and practice. Journal of Engineering Design. 9, 315–327 (1998).
Maher, M. Lou, Poon, J., Boulanger, S.: Formalising design exploration as co-evolution: a combined gene approach. In: Gero, J. (ed.) Advances in Formal Design Methods for CAD // Advances in formal design methods for CAD. pp. 3–30. Chapman & Hall, London (1996).
McAloone, T.C., Andreasen, M.M.: Design For Utility, Sustainability and Societal Virtues: Developing Product Service Systems. In: Marjanovic, D. (ed.) 8th International Design Conference—Design 2004. Design Society, Glasgow (2004).
Meißner, M., Gericke, K., Gries, B., Blessing, L.: Eine adaptive Produktentwicklungsmethodik als Beitrag zur Prozessgestaltung in der Produktentwicklung. Tagungsband 16. Symposium "Design for X". pp. 67–76 (2005).
Miriam-Websters: No Title. https://www.merriam-webster.com/dictionary/context.
Möhringer, S.: Entwicklungsmethodik für mechatronische Systeme, (2004).
Nicolescu, B.: Transdisciplinarity—Past, Present and Future. Congresso Mundial de Transdisciplinaridade. pp. 1–24. Vila Velha, Brazil (2005).
Pahl, G., Beitz, W., Feldhusen, J., Grote, K.H.: Engineering Design—A Systematic Approach. Springer-Verlag, Berlin (2007).
Roozenburg, N., Cross, N.: Models of the design process: integrating across the disciplines. Design Studies. 12, 215–220 (1991).
Roozenburg, N.F.M., Eekels, J.: Product Design: Fundamentals and Methods. John Wiley & Sons, Chichester (1995).
Sakao, T., Lindahl, M.: Introduction to Product/Service-System Design. Springer, London (2009).
Shostack, G.L., Kingman-Brundage, J.: Service Design and Development. In: Congram, C.A. and Friedman, M.L. (eds.) Handbook of Service Marketing. New York (1990).
MMEP SIG: Modelling and Management of Engineering Processes Special Interest Group - home page. https://www-edc.eng.cam.ac.uk/mmep/.
Tomiyama, T., Gu, P., Jin, Y., Lutters, D., Kind, C., Kimura, F.: Design methodologies: Industrial and educational applications. CIRP Annals—Manufacturing Technology. 58, 543–565 (2009).
TU Braunschweig: GINA. https://www.gina-net.de/.
Wallace, K.: Tranferring Design Methods into Practice. In: Birkhofer, H. (ed.) The Future of Design Methodology. pp. 239–248. Springer-Verlag London Limited, London (2011).
Wynn, D., Clarkson, P.J.: Models of designing. In: Clarkson, P.J. and Eckert, C.M. (eds.) Design Process Improvement A review of current practice. pp. 34–59. London (2005).

Design as a Common Denominator

Tatjana Leblanc

Abstract Design practices nowadays tackle society's needs at every scale and level of complexity. They focus not only on objects of daily use, but also on user interfaces, systems, cities, landscapes, services and even organizations. Design problems are therefore correspondingly more complex than they used to be, calling for an increasingly broader knowledge base and greater expertise. Using the case study of urban infrastructures and their impact on public space, this paper will illustrate the evolved nature of design and the challenges it encounters in its role as facilitator and convergence point of collaborative approaches. The paper will highlight the complex dynamics between the players involved as well as their roles; the logic guiding certain decision processes; and the challenges posed by the persistent misconception of design as a 'discipline.' Lastly, the dysfunctional aspects of collaborative approaches will be identified and recommendations for design education and practice, proposed.

Keywords Evolved design · Transdisciplinarity · Multidisciplinary collaboration

1 Introduction

The face of urban North America is changing and the utility distribution infrastructures contribute to that phenomenon (Poullaouec-Gonidec et al. 2005; Buchard 2002; Bekowitz et al. 2003; Gauthier 2006; Leblanc et al. 2008; Montpetit et al. 2002). Over the decades, these infrastructures have come to host a number of services on a shared support system known as the utility pole. As these poles with their contraptions and cables have proliferated and grown, managing the array of services they host has become increasingly difficult. The more elements are added, the more physical and visual saturation becomes a nuisance (Buchard 2002; Gauthier 2006; Leblanc et al. 2008).

T. Leblanc (✉)
École de Design, Faculté de l'Aménagement, Université de Montréal, Montréal, Canada
e-mail: t.leblanc@umontreal.ca

L. Blessing et al. (eds.), *The Future of Transdisciplinary Design*,
https://doi.org/10.1007/978-3-319-06382-9_2

These trends seem paradoxical in the age of nanotechnology, where ultra-compact, highly sophisticated devices are ubiquitous. Indeed, the undesirable impact of the growing utility network on the urban landscape has intensified debate on the notions of thresholds, social acceptability and quality living space (Montpetit et al. 2002), suggesting the need for a shift in priorities.

In this paper, we will use the case of urban utility infrastructures to illustrate the dynamics between the various players involved and point out the dysfunctional aspects of multidisciplinary approaches to problem solving. The Chair of Landscape and Environmental Design at the University of Montreal (CPEUM) has been for several years examining urban infrastructures and their impact on landscape and the quality of living spaces. Empirical and phenomenological research allowed the CPEUM team to look at this topic from multiple disciplinary angles and to compile data that vividly portrays their evolution and associated challenges (Poullaouec-Gonidec et al. 2005; Gauthier 2006; Leblanc et al. 2008; Montpetit et al. 2002). This research reveals not only the increasingly complex nature of design problems, the need for integrated knowledge and multidisciplinary expertise, but more so the challenging process of multidisciplinary actions. In identifying the shortcomings from a design perspective and elaborating on the role of design, the paper will draw a comparison between transdisciplinarity and design thinking.

Lastly, the concluding remarks will present the author's position with respect to the future of transdisciplinary design, design education and the role of academic institutions.

2 Case Study: Urban Development and Aerial Infrastructures

In recent years, many concerns have been raised about the quality of living environments and public space in urban settings (Montpetit et al. 2002). Various infrastructures dominate the urban landscape and define the use of public space: paved roadways, traffic signs, street lighting, parking areas, gas stations and the like. In addition, the sidewalks are lined with street furniture, including phone booths, lights, parking signs, billboards, parking meters, bus shelters, benches, water fountains, trashcans, mailboxes and bike racks.

Environmental concerns lead naturally to alternative modes of transport like electric vehicles and bike rental systems. Yet these too require specific infrastructures: charging stations, bike docking stations, bike paths, regulatory signage and so on. Judging from the outcomes, many of these features are implemented with little concern for visual identity or their impact on the surrounding environment.

Fig. 1 Examples of saturation and the coexistence of old and new structures

2.1 The Evolution of the Aerial Distribution System

The aerial distribution system has also become increasingly present and appears almost as something from another era. An engineering phenomenon comprised of wooden poles, electrical equipment and an arsenal of cables, it claims its space in the urban landscape, growing progressively more unwieldy as it struggles to deal with ever-increasing demands. Clearly, the rising demand in communication services has done nothing to assuage the proliferation of cables and apparatuses, as witnessed by the physical and visual saturation in Fig. 1. Today, many perceive these poles and their contraptions as imposing, disturbing and potentially dangerous (Leblanc et al. 2008; Montpetit et al. 2002).

Studies confirm that these awkward structures have raised public concerns and prompted questions on the threshold of social acceptability (Montpetit et al. 2002). Where are the limits? Are there alternative solutions? As shown in Fig. 3, design solutions are possible when the willingness exists and priorities are well assessed. Figures 2 and 3 illustrate such a difference.

2.2 The Contextual Environment

Many European cities implant their utility services in underground conduits. One may therefore ask, why such options are not considered in North America? In fact, the North American context is somewhat more complex. The simple size of the territories and extreme climate conditions are not comparable with the contextual situation of many European cities. Although many strategic or "marketable" locations benefit from the underground solutions, North American utility providers avoid these

Fig. 2 Engineering-driven solution

Fig. 3 Design-driven solution

options for several reasons: cost, inaccessibility of conduits during winter season (ice and snow accumulation), lack of flexibility (the need for permits to access or modify the system), and the disruption of traffic during maintenance and repair (Montpetit et al. 2002).

The evolution of infrastructures is a dynamic phenomenon driven by a number of factors, especially by the interconnectedness of stakeholders who occupy distinct parts of the system. The intertwined relationships among the constitutive parts of the actors-system-context problem field define its complex nature:

- Service providers: electricity, telecommunication, cable services, their infrastructures and technical components, etc.
- Stakeholders: government, municipalities, service providers/private partners, commerce, real-estate developers, private landowners and citizens.
- Other public equipment: urban furniture, lighting systems, signage systems, bike sharing systems, fire hydrants, drinking fountains.
- Technical concerns: efficiency, quality of service, functionality;
- Economic concerns: profitability, marketing, property value, cost, etc.
- Social and environmental concerns: security, public health and safety, natural territory.
- Aesthetic and cultural concerns: measures of mitigation, preservation, development, patrimony, landscape, and quality of living environment.
- Contextual environment: rural and urban areas and their distinct use (residential, commercial, industrial, recreational).
- Socio-political and socio-economic context.

Any change to the system affects in some way other stakeholders. To improve efficiency or modernize services, for example, utility providers have added new features and components, which today manifest themselves in the form of automated control devices and heftier cables. The existing poles have consequently become inadequate and need to be replaced with sturdier ones. However, due to the lack of coordination, the old and new poles continue to coexist (sometimes for several years), until the responsible parties find the time and resources to transfer their components onto the new structure (Fig. 1).

2.3 *Problem Solving and the Dynamics of Multidisciplinary Teams*

A number of experts and regulatory bodies—local authorities, urban designers, economists, architects, landscape architects, engineers, industrial designers, maintenance staff, unions and so on—are involved at some point or other in the design, development, implementation and maintenance processes. To deal with the complexity of managing both the infrastructure and the players involved, convoluted procedures have been instated, obviously driven by disciplinary expertise. These procedures determine the stages and scope of any disciplinary interventions. For example, urban planners frame the project and suggest trajectories, engineers determine the needs and establish technical parameters, economist elaborate the budgetary aspects, and so on.

Because utility infrastructures are implemented in the public domain, municipalities play the governing role by way of consultation processes. Their decision-making follows meticulous administrative procedures and guidelines (Hydro-Québec 1999, 2002) that are often based on, what some call, "best practices"—in other words; on those "that exist" and not that "ought to be". As such, they are motivated by past experiences and—what some seem to ignore—their own contextual environment.

One of the major problems surrounding this process is the lack of counter-expertise and interdisciplinary evaluation of the proposed design solutions, especially in light of the intertwined nature of the system and actors. In absence of counter-expertise, engineering- or cost-driven proposals are rarely challenged.

In fact, people believe that since utilities are essential, the consequences of their existence must be borne, no matter what shapes and sizes these may take. Those who argue in favour of underground solutions remain powerless in the face of the imposed economic and technical constraints of the current context.

3 The Shortcomings of Disciplinarity

To manage such complex, intertwined systems requires more than the makeshift approaches favoured by the authorities in the face of the abovementioned constraints. Approaches of this kind tend to treat the symptoms and not the underlying causes of a given problem (Leblanc et al. 2008). Once problems appear, authorities subcontract expert studies of impact analysis. Yet these only generate further guidelines and conventions and never provide comprehensive solutions to the problem at hand. Such practices seek mainly answers concerning methods, time frame and cost, in order to compare and assess new project proposals. However they miss to acknowledge the obvious: the changing context (needs, stakes, challenges, etc.). As Cross points out, "[m]ethod may be vital to the practice of science (where it validates the results), but not to the practice of design (where results do not have to be repeatable, and, in most cases, must not be repeated, or copied)" (Cross 2001).

Disciplinary expertise is useful when it relates to and enhances the understanding of complex phenomena. Nonetheless, it should do more than serve as a contributor: disciplinary expertise must also be an integral part of a design process that accepts alternative approaches and outside-the-box-thinking. As Einstein once said: "We cannot solve our problems with the same thinking we used when we created them" (Harris).

In the case of utility infrastructures, however, studies have shown that design solutions are only considered when engineering solutions have failed and corrective measures are needed. At this late stage of the process, design interventions are limited to providing a cosmetic fix. The fact that current disciplinary and multidisciplinary approaches tend to overlook such crucial concerns as impact on the environment and heritage, quality of living space and depreciation of property value clearly indicates the shortcomings of disciplinary thinking.

Treating urban development as a design problem would not only give all stakeholders a better grasp of the context and its inherent opportunities, but it would also give rise to alternative solutions capable of preventing unwanted ramifications.

Some have understood that "by working across different areas of expertise, strategic design outlines the 'architecture of the problem,' highlighting key opportunities for improvement in all aspects and outcomes of a problem" (Helsinki Design Lab).

As some might point out, there are examples of urban development that successfully implement alternative urban design perspectives and human-centered approaches such as the case of Freiburg im Breisgau (Baden-Württemberg, Germany), the Collingwood Village (Vancouver, Brittish-Columbia, Canada), or the Rosslyn-Ballston (Arlington County, Virginia, USA) (Vivre en Ville 2013). They all place a strong emphasis on the human scale and services while thoughtfully integrating the necessary infrastructures within the urban landscape. This begs the question: Why do these examples inspire so little North-American cities? The answers we get are often: The context is incomparable. Procedures and regulations differ. The scale of the problem is unparalleled, especially with so many actors involved. Yet, more importantly, what many avoid admitting, there is a strong resistance to change. The monopole position of service providers and the procedures that govern all urban projects are major hurdles that tend to obstruct innovative design alternatives.

Innovation does not always mean scrapping the existing and starting from scratch. Innovative and smart solutions are possible at any level, as shown in Fig. 4. The presented concept is the result of a transdisciplinary design approach and it illustrates a designer's ability to compose with complexity, challenges and constraints while remaining context-sensitive and attentive to users' needs. Several disciplines were involved in (engineering, urban planning, landscape architecture) in modelling the problem situation thus helping the designer to understanding the context and constraints, interpret their concerns, and materialize an alternative concept. This concept may not be *the* answer to the problem. However, to evaluate its real potential would require an open mind and the willingness to consider alternative ways of doing. Yet, from behind the shield of the complexity and cost arguments, stakeholders show resistance to change and are sceptical about ideas that threaten the status quo.

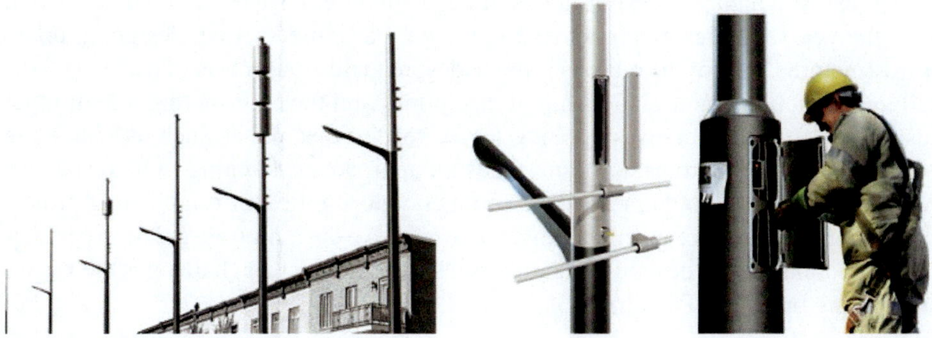

Fig. 4 Utility pole concept, housing components, with smart-detach system calibrated for weight overload during extreme weather conditions (student project 2008)

4 The Ambiguous Nature of Design

Design means different things to different people. During the industrial era, design was primarily concerned with the aesthetic qualities of industrial products. As such, it was tied to the economic context and seen as an engineering and marketing tool. This legacy, it seems, is hard to shake. People confuse design nowadays with arts and crafts; others think of fashion, graphics, styling and décor. Some use the term 'design' in reference to their professional activities: software development, web development, engineering, architecture, interiors, marketing, business development, urban planning, etc. This widespread use of the term has progressively diluted the meaning of design and helped bring about a certain ambiguity.

As times have changed, so has design. During the 1960s, progressive design thinkers introduced the notion of design as a *science of the artificial* (Simon 1981), thus laying the groundwork for a major paradigm shift. This triggered debates about design, its methods and the designer's role in defining the future of society (Simon 1981; Alexander 1964; Gregory 1966; Schön 1983). The result was a reshaped identity along with the transformation of design practices and education.

Design has since matured into a reflective practice (Jonas 1999) that places the user and societal needs front and centre. Not only does it focus on people's perceptions and their experience of the material world (Krippendorff 2006), it also pays particular heed to such modern-day social concerns as demographic shifts, environmental impact, and social problems.

To tackle complex issues, designers have learned to collaborate and developed cross-cultural skills. They learned to transcend disciplinary frontiers and interpret expert knowledge into meaningful design solutions (Cross 2001; Simon 1981; Gregory 1966; Schön 1983; Pinson; Jonas 1999). They thrive in cross-disciplinary environments, which feed their imagination. Transdisciplinary thinking integral to design thinking is a quality that nowadays defines design and justifies its role as facilitator, integrator and interpreter (Jonas 1999).

The evolved nature of design and its strategic qualities have been well documented over the years, yet design continues to play only a limited role in designing urban infrastructures. One of the reasons is the widespread misconception of design and the failure to teach the interrelationship of disciplines and the contributing role of other disciplines. It is especially surprising to see related disciplines such as marketing or engineering resist transdisciplinary design approaches, clinging to its respective roles in identifying market needs and product opportunities, or establishing project parameters and product specifications. Some perceive design's transgression of disciplinary boundaries as destabilizing, others, as an intrusion that challenges their ways of thinking or doing (Pinson).

5 Transdisciplinarity: Feasible or Utopian?

Transdisciplinarity can be interpreted as one's ability to look at problems from various perspectives without being limited by disciplinary ideology, conventions or principles. Scholars describe it as a "state of mind that refuses a singular discourse" and recognizes "the existence of different levels of reality" (Nicolescu 2002). This implies recognizing the interdependency of disciplines and their relationships to real life, accepting other views as an enrichment of one's own and learning to benefit from this insight in pursuing the best possible solution (Nicolescu 2002). Morin concurred "monodisciplinarity or isolated knowledge leads only to a 'blind intelligence'. Knowing is an uninterrupted loop of separating in order to analyze and reconnecting in order to synthesize." He explains "by compartmentalizing knowledge, people lose their ability to contextualize and reposition the knowledge in its natural context" (Morin).

However many insist that bringing disciplines together is not enough; they need to **merge their expertise**, build a **common understanding** of the problems at hand and develop **shared strategies** for resolving them (Morin).

Scholars agree that disciplinarity complements transdisciplinarity. Disciplinarity offers a deeper insight into a fragment that is part of something larger. All fragments of knowledge, however, need to be merged and arranged into a meaningful *whole*. This cognitive effort is considered transdisciplinary thinking. It represents a true paradigm shift, since "it is concerned with what is at once in between, across and beyond all disciplines" (Nicolescu 2002). As Papst suggests, as a cognitive function, transdisciplinarity is an **aptitude that can only be developed by the individual mind**. Understandably, such aptitudes need to be developed since they are key to fertile multidisciplinary collaboration (Papst 2003).

Transdisciplinarity should not be confused with multidisciplinary approaches, which lead to an accumulation of expert opinions and rarely produce a satisfactory outcome. They lack a unified view of the problem. In such a context, team dynamics can often be characterized as a show of expertise instead of a pooling of knowledge. A number of challenges inhibit transdisciplinary thinking and contribute to a dysfunctional climate. The first missteps occur during the start-up phases of a project, which call for a framework (objectives, scope, budget, schedule, etc.). These challenges typically trigger questions, such as: Who does what? Who initiates the project? Who is in charge? Who defines criteria and product specifications? Such questions reveal a strong preoccupation for hierarchy and procedures rather than understanding the problem, defining common goals and developing strategies for attaining them.

Needless to say, cultural differences and disciplinary boundaries can be an obstacle to transdisciplinarity. They tend to intimidate, prompting defensive attitudes and concerns for keeping the balance of hierarchical power. Particularly problematic is reconciling data-driven empiricism (which leaves little room for doubt) with design thinking, which is all about finding new ways of seeing and doing. Such inherent differences of disciplinary cultures can indeed inhibit transdisciplinary thinking. As some scholars point out, design approaches differ to scientific methods in that the former are concerned with defining the future and anticipating needs, which has no precedent, while the latter tend to focus on past and present phenomena (Vivre en Ville 2013; Simon 1981; Schön 1983).

Designers are used to questioning existing solutions and challenging traditions and conventions. They realize that today's answers may not be appropriate for tomorrow since, given the perpetual evolution in perceptions, behaviours and needs, the future is highly uncertain.

Yet not all disciplines nurture such a thought process. Those familiar with design culture show little resistance to transdisciplinarity because they learned to value complementarity; conversely, those who are unfamiliar remain sceptical, self-protective and bent on maintaining their ways of thinking and acting.

6 Conclusion

This paper set out to draw attention to the current situation of urban infrastructures and stress the need for a paradigm shift. As they currently stand, North American urban infrastructures show serious design flaws, having become an uncontrolled support system for a vast range of services and devices. Technical devices multiply and cables increase not only in number but also in size, despite intensified public pleas to protect the quality of urban environments (Poullaouec-Gonidec et al. 2005; Montpetit et al. 2002).

Future developments will need to be aware of the complex nature of the urban phenomena as well as deal with emerging issues like quality living space, heritage management, visual, sonic and atmospheric pollution, urban ecology and public health, and sustainable lifestyles (Gauthier 2006). As indicated, transdisciplinary design approaches can lead to sustainable and socially acceptable alternatives.

The case study made the dysfunctional dynamics of multidisciplinary teams apparent. The identified problems can be attributed to difficulties in merging disciplinary cultures and breaking with conventional disciplinary thinking, formed through educational systems. This, in turn, illustrates the need for better communication about design and transdisciplinary thinking. To be effective, disciplines will need to develop tolerance and welcome creative ideas with an open mind.

As many point out, transdisciplinarity will be a key skill for future professionals and scientists to bring to the table (Morin), emphasizing the need for academic institutions to adapt their programs and teaching methods accordingly. Nurturing the development of a wider set of skills—critical and creative thinking, analysis capabilities, new-media literacy and the ability to work in multi-disciplinary settings with a transdisciplinary attitude—will be crucial. Academic institutions will thus need to establish interdisciplinary learning environments (Leblanc 2009) where students can develop such skills while being exposed to a wide range of subjects (IFTF Future Work Skills 2020). In this process, "[w]e should teach methods of grasping mutual relations and reciprocal influences between parts and the whole in a complex world" (Morin). However, to teach transdisciplinary approaches, academic institutions and faculties will too need to overcome disciplinary boundaries. For instance, students from the Harvard Graduate School of Design challenge their Harvard community: "Can we unite Harvard through design... lead the movement to break down the vestigial silos that separate students, methodologies and reactive ideas between the schools?" (IFTF Future Work Skills 2020).

Progressive voices of design characterize such initiatives as follows: "For us it's the conscious design that builds political, economic, and social interests towards a desired state; it's the art of getting there. And when we speak about design in this context, we speak of it as a leadership model: a way of leading in an uncertain world, where iteration is the key to connecting opportunity to impact. In our work, design and stewardship are interconnected" (xDesign Conference 2013). As such, design is the common denominator of all disciplinary efforts that shape the future of societies.

Acknowledgements The research on urban infrastructures was conducted in association with the interdisciplinary research team of the CPEUM, la Chaire en Paysage et Environnement de l'Université de Montréal. Special thanks go to my research assistant Yolaine Turcotte, who developed the design concept during her thesis project (Fig. 4). All other images used in this document are the author's.

References

Alexander, C.: Notes on the Synthesis of Form. Harvard University Press, Cambridge (1964)

Bekowitz, A. R., Nilon C. H., Hollweg K. S.: Understanding Urban Ecosystems. A New Frontier for Science and Education. Springer-Verlag Inc., New York (2003)

Buchard, S.: Être beau. Le Devoir, October 1, (2002), https://www.ledevoir.com/non-classe/10233/etre-beau

Cross, N.: Designerly Ways of Knowing: Design Discipline Versus Design Science. Design Studies, 17(3), 49–55 (2001)

Gauthier, M.: La ville, l'urbain et le développement durable. Natures sciences sociétés: rétrospectives et prospectives. Natures sciences sociétés, 14(4), 383–391 (2006)

Gregory, S. A.: A Design Science. In: S. A. Gregory (ed.) The Design Method. Butterworth Press, London (1966)

Harris, K.: Collected quotes from Albert Einstein, https://rescomp.stanford.edu/~cheshire/EinsteinQuotes.html

Hydro-Québec: Lotissement et réseaux de distribution souterrains : guide des bonnes pratiques. In association with Bell Canada, Vidéotron and the Union des municipalités du Québec (2002)

Hydro-Québec: Lottissements et réseaux aériens: guide des bonnes pratiques. In association with Bell Canada and the Union des municipalités du Québec (1999)

IFTF Future Work Skills (2020), https://www.iftf.org/our-work/global-landscape/work/future-work-skills-2020

Jonas, W.: On the Foundation of a 'Science of the Artificial.' In: HEL99 International Conference, UIAH Helsinki (1999)

Krippendorff, K.: The Semantic Turn—A New Foundation for Design. Taylor & Francis, Boca Raton/London/New York (2006)

Helsinki Design Lab, https://www.helsinkidesignlab.org

Leblanc, T. et al.: Élaboration de critères de la capacité d'accueil des poteaux de distribution en milieu urbain et redesign du boîtier de commande d'automatisation. In: Rapport de la Chaire en paysage et environnement de l'Université de Montréal (2008)

Leblanc, T.: Transdisciplinary Design Approach. In Creativity and HCI: From Experience to Design in Education (pp. 106–122). Springer US, (2009)

Montpetit, C., Poullaouec-Gonidec, P., Saumier, G.: Paysage et cadre de vie au Québec: réflexion sur une demande sociale émergente et plurielle. Cahiers de géographie du Québec, 46(128), 165–189 (2002)

Morin, E.: Quelle Université pour demain? Vers une évolution transdisciplinaire de l'Université, https://ciret-transdisciplinarity.org/bulletin/b12c1.php (author's translation)

Nicolescu, B.: Manifesto of Transdisciplinarity. State University of New York (SUNY) Press, New York. Translation into English by Karen-Claire Voss (2002)

Papst, J. (Graz): From signs to the perspective of beyondness or transdisciplinarity. In TRANS. Internet-Zeitschrift für Kulturwissenschaften. No. 15 (2003), https://www.inst.at/trans/15Nr/01_6/papst_b_15.htm

Pinson, D.: L'Urbanisme: une discipline indisciplinée. https://hal.archives-ouvertes.fr/hal-00789207

Schön, D.: The Reflective Practitioner. Bibliography review 63. MIT Press, Cambridge (1983)

Simon, H. A.: The Sciences of the Artificial. 2nd edition. MIT Press, Cambridge (1981)

Vivre en Ville: Retisser la Ville: [Ré]articular Urbanisation, Densification et Transport en commun (Volume 1). Retisser la Ville: Leçons de Cinq Experiences de Transit-Oriented Development (Volume 2). Collection "Outiller le Québec" (2013)

Poullaouec-Gonidec, P., Domon, G. and Paquette, S. (dir.): Paysages en perspective. Les Presses de l'Université de Montréal, Montreal (2005)

xDesign Conference (2013), https://harvardxdesign.com/dossier

Ineffective Collaboration in Multi-Disciplinary Teams

Angela Fernandez-Orviz

Abstract Our traditional approaches to problem solving and innovation are obsolete. The growing complexity of our needs and the problems that our life styles generate, demonstrate the increasing need for a more holistic approach which integrates the knowledge of all relevant areas from the early stages of a project. Understanding the complexities of these collaborations is the first step to build the appropriate frameworks that enhance their success. This paper presents a comprehensive classification of the most common challenges encountered in multi-disciplinary collaborations and argues for the value of making team members aware of these potential issues with the prospect of minimising their effects. Finally it is discussed the potential suitability of Service Design techniques and processes to become the 'enabler' of these holistic approaches.

Keywords Multi-disciplinary · Collaboration · Design

1 Introduction

The increasing complexity of the challenges faced by our society is exposing the need for a more exhaustive evaluation of all the aspects involved and thus has instigated a growing demand on understanding holistic and collaborative approaches and establishing the basis for their effectiveness.

The value in these collaborations is here understood to rely on a higher integration of the relevant knowledge providing a non-hierarchical approach to knowledge in order to achieve the appropriate solutions and empower innovation through both creativity and expertise (Souter and Billout 2007, p. 136). This approach to collaboration builds on Hollins' definition of 'design circles': spaces which enable "people with different expertise and backgrounds to meet on an equal footing and communicate effectively with each other" (Hollins and Hollins 1991).

A. Fernandez-Orviz (✉)
Edinburgh College or Art, University of Edinburgh, Edinburgh, UK
e-mail: a.orviz@ed.ac.uk

L. Blessing et al. (eds.), *The Future of Transdisciplinary Design*,
https://doi.org/10.1007/978-3-319-06382-9_3

In theory, multi-disciplinary teams' have the augmented capability of simultaneously providing specialised knowledge for focusing on the detail, together with a greater collective understanding of the subject. This duality enables identifying weaknesses that would not be appreciated when analysing its different aspects independently. These benefits are here attributed to two main characteristics:

- Gathering different perspectives helps in breaking assumptions and inspiring new ideas.
- The balance provided by the different expertise supports the work towards feasible solutions.

However, the complexities inherent to teamwork, which are here speculated to be emphasised by the diversity present in multidisciplinary teams, often overshadow the potential of these collaborations. The increased specialization of our educative systems is here proposed to be at the core of these challenges, having two main negative effects in the development of successful collaborations between different disciplines.

On the one hand, experts have a limited appreciation of factors that are external to their discipline and therefore tend to lose perspective with respect of the bigger picture (Rees 2010). This issue has been eloquently exposed by Fung et al. (2006) by saying: "When the only tool you have is a hammer, every problem looks like a nail". As defined by Robert J. Stenberg (Sternberg 1994), "expert performance reflects an adaptation to a specific domain to attain high performance on tasks within that domain". That adaptation to certain tools and approaches provides a narrowed vision of the context and potential solutions for a problem.

On the other hand, the specific methodologies and terminologies developed to enable communication between professionals of the same field, plays against the understanding between professionals from different disciplines. It is therefore essential for effective collaborations to build a common language beyond the specification of each discipline.

This need for adaptation to collaborative environments is already having an impact on our educational systems, and more often courses that foster these approaches by gathering students from different disciplines are being developed. This essay proposes the value of comprehensively outlining common challenges as part of that training.

Therefore, this essay is concerned with exposing the challenges identified as common in this type of collaborations and the exploration of their roots and effects. Also, it will be proposed the potential suitability of taking a Service Design approach as the means for mediating and enabling these collaborations, building towards a new definition of Trans-disciplinary Design.

2 Terminology Approach

Regarding the terminology debate, a variety of prefixes are often added to the term 'disciplinary' in order to describe a variety of approaches and collaborations.

These prefixes carry themselves a semantic meaning (Oxford 2012): intra- (within), inter- ("between; among"), multi- ("more than one; many"), cross- ("denoting movement or position across something" either "denoting interaction" or "passing from side to side") and trans- ("across; beyond").

But the meaning of these built terms is also dependent on their contextual use, of which four different purposes have been identified:

1. The topics to be tackled by a project and their domain
2. Individuals and their fields of expertise
3. The fields of expertise covered by the team as a whole
4. And ultimately the approach, which is sometimes defined according to the expertise held by the individual/team and the topic to be tackled, and other times depending on the "level of integration" of the different fields involved.

Besselaar and Heimeriks (2001) claim that these "forms of non-disciplinary knowledge [...] are generally defined in contrast to what is seen as 'normal'" within the methodologies, procedures and fields of application of each discipline. But as our tools and knowledge evolve, especially with the development of new technologies, so to does the designation of what 'normal' is. It could be said that the definitions and boundaries of our disciplines evolve in a Darwinian manner. There are constant little modifications, 'experiments', and those that succeed stay and become either part of the discipline or a new subdiscipline. And in that manner non-disciplinary approaches become intra-disciplinary.

The variety of interpretations and the ephemeral character of their definition may make it impossible to reach a nomenclature for these collaborative and mixed approaches that is unified and consistent over time. Therefore, without entering further this debate, the use of this terminology will be defined within the context of this essay. In this research, the emphasis is put on collaboration. Therefore, these collaborative groups of experts from different domains will be referred to as multi-disciplinary teams. On the other hand, the approach is to be holistic and the knowledge of those experts is to be integrated beyond the boundaries of any of their disciplines, thus it will be here defined as a trans-disciplinary approach.

3 Rationale

The author's experience within multi-disciplinary teams revealed a series of challenges that triggered this work back in 2009, as part of the Masters in Design Innovation at the Glasgow School of Art. Since then the author has kept on working in multi-disciplinary, cross-institutional and cross-sector projects, which have brought new insights on the matter.

The initial research, mainly done through literature review, was focused on establishing the benefits and meaning of multi-disciplinary collaborations as well as the stages of a project in which these collaborations acquire more relevance. Although that work will not be here presented, its conclusions provide the rationale for the definition of collaboration offered in this essay. It established a relevant difference between working collaboratively and working in a well-coordinated manner. That differentiation relies on the way knowledge is applied. Collaboration is in this context defined as the integration of knowledge from different fields in order to build new knowledge or solutions. So it will be referring to collaborative thinking as the means for problem solving and innovation. Whereas working in a well-coordinated manner would simply imply using the expertise of different fields in order to develop different aspects of a project, which is a very different and less holistic application of that knowledge (Hollins and Hollins 1991, p. 137) (Buchanan and Huczynski 1997, pp. 1–14). Therefore, as collaboration is the process of 'thinking together', its highest value is attributed to the early stages of a project—those of research and conceptualization.

The work presented here was driven by the belief that there is value in informing the team members about the potential issues that may challenge the success of their collaboration. It is proposed that providing them with a comprehensive overview on the potential issues present in these contexts is the first step to prevent or minimise their effect by augmenting the teams' capabilities for collaboration through awareness. With this purpose a 10 min animated video[1] was created, establishing some of the essential aspects that challenge a successful collaboration. This video is currently being used as a source for training in a variety of contexts, from industry or online training to academic teaching. This reinforces the initial assumption on its value. Furthermore, it is proposed that future work building on the issues outlined here could develop appropriate frameworks and tools to improve the relationships and outcomes of these collaborations.

4 Research

The insights presented here have been drawn through the analysis of specific situations encountered in a variety of multidisciplinary contexts, complemented with literature review and semi-structured interviews with professionals working in interdisciplinary environments.

Some of the projects used as case studies involved students of the Masters in Design Innovation at the Glasgow School of Art, working in real contexts with communities and public services providers in the Glasgow area. The most relevant of these projects the GetGo Glagow initiative, winner of the Audi's competition Sustain Our Nation through which working in collaboration with a variety of stakeholders and the community of Wyndford (a neighborhood in Glasgow) the social enterprise 'Green Gorillas—now run by community members—was set up. This

[1]Effective Teamwork & Collaboration - https://www.youtube.com/watch?v=NsndhCQ5hRY.

group was formed by 12 students, including the author, from a variety of disciplines such as graphic, product, costume and service design, business, engineering, art history and photography. As an educational exercise the project was envisioned to last for four months. However, the students' commitment to the project and the success in the application for funding resulted in the extension of the students' work for longer than a year. Other shorter projects in collaboration with the South Lanarkshire Council were also delivered by the same group subdivided in smaller teams, such as the redevelopment of the Blantyre's Quarry or the South Lanarkshire Libraries Project in which the future and evolution of libraries was being explored. Certain aspects of team dynamics, such as differences between external and self-perception, were explored through a design-led approach by the development of activities for a feedback workshop undertaken with the Masters' students.

Other insights have been drawn from the authors' experience in entrepreneurship educational programmes where students and professionals from different fields are put into unmediated groups to develop concepts and prototypes for fictional scenarios under time pressure. Also some examples are from a fictional exercise concerned with the commercial and technical development of an island, involving 22 students from a variety of engineering disciplines and led by the author was used as a case study (2007).

Finally, the author's professional practice at Moving Targets[2] has provided her with insight into cross-institutional academic research involving practitioners from a variety of disciplines such as Human Computer Interaction, Digital Arts, Architecture, Economics, Marketing or Videogames Development; as well as experience working with a variety of creative sectors. Through this practice, some of the arguments developed with the initial research have been reinforced and some new insights have been gathered. Through the facilitation of creative workshops involving participants from different industrial sectors and academic background, the potential of developing tailored tools and activities that can mediate discussions and collaborations in real contexts has also begun to be explored.

Due to the interpretative character (Swan 2002) of the research and its limited scope, it is acknowledged that further difficulties may appear in the longer term or in unexplored contexts. However it is sensible to think that if the essential vulnerabilities of a multidisciplinary team are not covered from the beginning the cohesion of the team is more likely to break; preventing the success of the collaboration.

5 Barriers to Multi-Disciplinary Collaborations

Both general group dynamics and team building strategies have been widely studied in order to develop systems and approaches that can minimize the risks of collaboration. However, even assuming a team has been shaped by compatible mind-sets

[2]Moving Targets is a Knowledge Exchange project funded by the Scottish Funding Council through the Horizon Fund.

(following for instance Myers-Briggs Type Indicators) and all the essential roles are covered (as defined by Belbin), team dynamics are unpredictable. Besides when the team is formed by professionals from different backgrounds it is argued here that some of the common obstacles become accentuated and some new challenges may appear.

5.1 Communication Barriers

Although communicating effectively is a common challenge for teams, certain aspects of it become stressed within multidisciplinary teams by what will be here defined as 'language differences' (Bucciarelli 2002) and 'intrinsic assumptions'.

Both of them can contribute to the development of 'parallel conversations', concept identified by Argyris (1991) as two professionals talking to each other but not actually listening. In multidisciplinary teams these are not necessarily a symptom of a listening blockage but the presence of obstacles to the mutual understanding.

Ruete (2000) identifies a series of forms of "non-participation" such as interrupting, negativity, physically or mentally leaving or dominating a conversation. These are inherent challenges to any kind of team. However, some of them are emphasised by a particular facet of multidisciplinary teams: the 'language differences'. Each discipline uses its very specific terminology which enables practitioners from the same background to communicate efficiently and effectively. Unfortunately this plays against the mutual understanding within multidisciplinary teams.

One of the interviewees—Ann Marie Shillito,[3] researcher and entrepreneur working between the domains of jewellery design and software development—described the frustration generated in her initial collaborations with engineers and software developers by what in the end was simply a different understanding of the word 'tool'.

During the observation periods it was noted that many discussions and arguments were ended by the simple phrase "but what do you mean by…" followed by each participant's definition of the specific term or concept. It seems sensible to think this kind of misunderstandings could be prevented by establishing a 'common language'. Building on this, some strategies are beginning to be developed such as encouraging and training team members to translate disciplinary discourse into plain English as well as sharing a glossary of terminology 'frequently used' in their discipline.

Another aspect inherent to multidisciplinary teams that accentuates general communication barriers are the 'intrinsic assumptions'. Members of an intra-disciplinary team can always have differences of opinion that are hard to deal with, but they will be setting out from the same basis. However in a multidisciplinary team each member brings along a set of tools and approaches (Fung et al. 2006), which

[3]Co-founder of Anakik 3D, explores the incorporation of haptic devices and 3D printing within traditional arts and crafts making processes. https://www.anarkik3d.co.uk/.

entails a series of assumptions and systematic procedures intrinsic to their field of expertise.

Thomson (2009) puts forward (pp. 27–34) the value in gathering different perspectives by arguing that it is "more difficult to identify the assumptions underlying our own reasoning than to identify the assumptions upon which others are relying". However, putting this into practice has proven challenging, as each individual is at risk of becoming "an advocate for his or her own technical specialty" (Brown and Katz 2009). But several of the situations analysed proved that it is not necessary to disagree, for this intrinsic assumptions to present a problem. Reaching the same conclusion through different logics may be reason enough for professionals not to trust one another. Therefore still agreeing on the final outcome, the reasoning path may be endlessly under debate unless these conversations are mediated.

5.2 *Uncertain Roles and Contributions*

The results of any collaboration are directly dependent of the team members' engagement, commitment and motivation towards the project. As Neumeier reveals: "strong-willed people love to collaborate when there's a sharp delineation of roles, an unobstructed view of the goal, and a strong commitment to quality" (2009). The lack of transparency or understanding in terms of the aim of specific activities has also come up as a cause for disengagement in the teams observed, often materialised in the form of questions such as "Why are we doing this?" and seemed to be often triggered by lack of clarity and miscommunications.

Professor John Briggs[4]—one of the interviewees—who has led numerous collaborations, claims his teams' engagement and commitment to be a key reason for their success: "The people working on international development offer less resistance to share knowledge and collaborate because they all have a shared interest in solving poverty and a shared sense of injustice". But that commitment not only does not always come naturally, but can be undermined but certain aspects of collaboration.

Fear has been recognized by researchers such as Argyris (1991, p. 104) and Swann (1987, p. 1038) as an important barrier for communication and, I would add, a cause for disengagement. Fear to intervention and contribution can be produced or stressed by the multidisciplinary environment. Professionals are removed from their comfort zone—their traditional field of work—and introduced into a new environment richer in perspectives and approaches. The potential for experimenting, fear to the unsuitability of their expertise or to find their input undervalued is therefore increased. But techniques for collaborative thinking that deal with fear and inhibition already used in Design, such as the simultaneous contribution of thoughts mediated through writing and encouraging environments free of criticism, suggest a potential solution.

[4]Pf. John Briggs works for The Centre of International Development at Glasgow University. https://www.gla.ac.uk/about/internationalisation/whoweare/africa/.

In most of the student projects analysed, the lack of commitment of other members as well as the lack of definition regarding the roles and contributions to be played proved to be strong causes for disengagement. Although accentuated by the lack of structure in the educational context, these issues are not far from the professional practice, and will be further discussed in the next section.

Therefore, a dynamic but clear definition of individual and collective roles is beneficial for the good functioning of any team (Buchanan and Huczynski 1997) (Neumeier 2009). But the "lack of understanding of the roles that the contributing disciplines can play" has been identified as a common cause of failure in these collaborations, thus that definition should go beyond the individuals also comprising their fields (Lyall et al. 2009).

At the beginning of this essay it was argued for the value of taking a non-hierarchical approach to research and idea generation in order to enhance the benefits of merging several fields of expertise. However, it would be naive to believe that all disciplines will have the same weight at all stages of the process. A discipline may find its role "subordinated to the main domain", being used mostly as a tool and without much voice in the overall project (Lyall et al. 2009).

However, in a truly inter- or trans-disciplinary project the degrees of contribution of each field might be unknown at the beginning and therefore impossible to pre-define. This indeterminacy can lead professionals to have "unrealistic over-expectations" of their own contribution as well as a "trivialized view" of the contribution to be played by other disciplines. These situations can trigger frustration and undermine the commitment of team members, thus the importance of managing the team´s expectations. This suggests the need for building that uncertainty into the team's understanding and expectations of the project, making a special emphasis on the dynamism and potential changeability as the project evolves.

As already argued, professionals often find it difficult to acknowledge that a better solution can be reached following a different procedure or breaking some of their intrinsic assumptions. This sort of 'professional arrogance' that hampers communications and risks to dismiss relevant knowledge is built on previous experience where a specific solution or process may have seemed to work in a similar situation. As William Rees puts it (2010) "We [experts] may think we know a very great deal, and we do but it is such a tiny part of the overall puzzle that it doesn't contribute very much [on its own]". Building that understanding of collaboration and that humbleness into a team would facilitate communications and aid managing expectations regarding the specific contributions of each field.

In terms of motivation for academic environments, the ISSTI's report on interdisciplinary research groups (Lyall et al. 2009), hints to the importance of having in place a "publication strategy" attending to the individual agendas and career goals. The free flow of information is indispensable to the success of any collaboration. But interviewees claim that information is often 'traded' rather than openly shared. The openness required for collaboration is especially challenging when new knowledge is being generated or team members belong to independent institutions, and does not seem to be encouraged by the credit system through publication existent in academia.

5.3 *Discrepancy Between Self- Perception and External Perception*

The ways in which we behave and understand our actions do not always match the impressions we make on our colleagues. Previous studies (Swann 1987) reveal that the discrepancy between self-perception and external perception can cause significant misunderstandings, leading to the breakdown of good team dynamics.

Through observation and interviews with individuals in several of the projects used as case studies, fear was identified as a common reaction to both providing and being exposed to negative feedback. But not debating existing issues or entering un-mediated discussions proved to equally generate conflict and tension within the teams. Providing a low risk environment for regular feedback would seem like the ideal solution to explore those potential misunderstandings. However, openly discussing the teams' strengths and weaknesses in a constructive manner remains a challenge. Literature review and interviews, revealed that most of traditional debriefing and feedback techniques (360°) employed by companies are perceived as tedious or bring with them the constraints of face-to-face feedback.

In order to explore further these differences between self-perception and external perception and avoiding using existing feedback techniques, playful activities for a feedback workshop were designed. Ice-breakers and competition games rewarding the 'best team members' were used to gamify the workshop. But written storytelling is the mechanism used for actual feedback deployed through two different activities.

In the first one, each participant was provided with their own 'feedback book' with blank pages as well as supporting cards that gathered a variety of adjectives and phrases related to working and personal qualities, aiming to set the tone of the feedback. Initially, participants were briefed to fill the initial pages of their books by describing themselves and their work within the team. The rest of pages are to be filled by the rest of team members attending to the same brief but regarding their colleagues. This activity is done anonymously but simultaneously.

For the second activity, fictional stories were generated by the participants having one of their colleagues as protagonist in each of them. In order to drive insight into their team dynamics, three fictional scenarios related to behaviour under pressure, decision-making, interaction and problem-solving were provided. Keen participants were encouraged to read those stories out loud to generate a collective experience.

This workshop was undertaken with team members of GetGo[5]—a real project run by students in which the author had participated—and used as case study for analysing the differences between self and external perceptions. It was found that negative personal and working qualities—while unrecognised by the individual—were consistently spotted by the rest of team members. The uniformity in the external perception

[5]GetGo was a project run by students at the Glasgow School of Art (GSA) winner of Audi Sustain Our Nation Contest through which the Social Enterprise 'Green Gorillaz' was set up and is now run by the community of Wyndford, Glasgow.

https://www.designweek.co.uk/glasgow-students-win-audi-sustain-our-nation-contest/3010183.article https://www.getgoglasgow.co.uk/.

leads to the thought that the team as a whole is likely to be right, whether or not it matches the self-perception. Various reasons for that discrepancy are speculated.

1. Self-perception is just wrong
2. Self-perception is based on previous experience, which was not shared with the rest of the members, and therefore cannot inform their judgement.
3. The person's reasoning is not coming across, so people cannot understand what causes a specific action or behaviour.

Not having an understanding of our colleagues' professional practice makes it more difficult for us to empathise with their previous experiences or comprehend the reasoning behind their actions or decisions. Thus discrepancies triggered by the two last causes are likely to become accentuated in multidisciplinary teams.

It is relevant to highlight that this workshop was developed as the means of research for deepening the understanding regarding self- and external perception. However, it is believed that these types of activities deployed on a regular basis can help build trust and strengthen the team's relationships. The potential for building a framework for safe feedback by using playful activities triggered a new line of research. Further work exploring these lines is expected to begin in 2014.

6 Trans-Disciplinary Design

The requirement for external mediation has been suggested several times throughout the exposition of the issues identified. It is now proposed how the Service Design methodologies could become a suitable approach for this multidisciplinary collaborations building a consistent framework for Trans-disciplinary Design.

In this context Trans-disciplinary Design is understood as an open approach in which professionals from different fields (including design disciplines or not) gather together in a project to collaboratively explore how their different expertise and research techniques can better merge to conceptualise and design more suitable solutions to complex problems.

The young discipline of Service Design was initially conceived through the development of visual tools that could enable the mapping and analysis of the different agents and relationships involved in the delivery of a service. However, in the past years the discipline and its approaches have evolved from user-centred to user-driven by exploring co-design and co-creation and building on more participatory techniques. In its approach, Service Design focuses on gathering and connecting the expertise and tacit knowledge collected by the experts—the people that use, deliver and manage the services. These are multidisciplinary groups of people which in many occasions present a variety of socio-cultural and economic complexities, and in which mutual understanding benefits from the 'enabling' role played by the discipline. Therefore, Service Design does not contribute to this design process with the expertise used in designing better solutions, but with the structure to it.

With this I would like to point out the similarities in context and execution between Trans-disciplinary Design and the Co-design approaches used in Service Design. Multidisciplinary collaborations could benefit from the experience already gathered being the starting point for developing the appropriate tools and processes. Our understanding of Trans-disciplinary Design can now move from the collaboration itself to the means through which it happens. It is the approach and processes through which all relevant knowledge is gathered, structured and glued in order to generate a new and holistic vision that enables the development of better solutions.

7 Conclusions

The potential for building appropriate procedures for the success of these collaborations building on the issues here identified was proposed at the beginning of this essay. In summary, this proposition could be simplified as follows:

- Communication barriers: rooted on 'language differences' and 'intrinsic assumptions', would benefit from enhancing a common understanding. It is proposed that building a basic understanding of each of the fields involved as well as having those discussions mediated by an external entity that can balance forces.
- Disengagement: Transparency, consideration of institutional and individuals' agendas, and the use of design methods for dealing with fear to contribution are proposed as the means to escape predictable disengagement.
- Uncertain Roles and Contributions: Their effect could be minimised by ensuring a 'dynamic definition' and building 'uncertainty' into the team's expectations. However, dealing with 'professional arrogance' might require a deeper change to our educational systems.
- Differences between self-perception and external perception: could be minimised by providing a low risk environment for frequent feedback.

The implementation of these ideas into proven tools and guidelines that can be put into practice is still an on-going research. However, through this essay a comprehensive classification of the potential issues that challenge the success of collaboration is put forward. Raising awareness within the team about these potential issues is proposed as the first step to prevent or minimise their effect.

References

Argyris, C., 1991. Teaching smart people how to learn. *Harvard Business Review,* May-June.

Besselaar, P. V. D. & Heimeriks, G., 2001. *Disciplinary, Multidisciplinary, Interdisciplinary—Concepts and Indicators*—Sydney, s.n.

Brown, T. & Katz, B., 2009. *Change by Design.* 1st ed. New York: Harpercollins.

Bucciarelli, L. L., 2002. Between thought and object in engineering design. *Design Studies,* 23(3), pp. 219–231.

Buchanan, D. & Huczynski, A., 1997. *Organizational behaviour, an introductory text.* 3rd ed. Milan: Prenice Hall Europe.

Fung, A., Lo, A. & Rao, M., 2006. *Creative Tools.* Hong Kong: Hong Kong: School of Design, The Hong Kong Polytechnic University.

Hollins, G. & Hollins, B., 1991. *Total Design: managing the design process in the service sector.* London: Pitman Publishing.

Lyall, C., Williams, R. & Meagher, L., 2009. *A Short Guide to Developing Interdisciplinary Strategies for Research Groups,* s.l.: ISSTI Briefing Note.

Neumeier, M., 2009. *The Designful Company.* Berkeley: New Riders

Oxford, 2012. *Oxford Dictionary Online.* [Online] Available at: https://oxforddictionaries.com. [Accessed 8 August 2012].

Rees, W., 2010. *Interdisciplinary education needed to address complex issues* [Interview] (18 March 2010).

Ruete, E. S., 2000. *International Association of Facilitation.* [Online] Available at: https://www.iaf-world.org/i4a/pages/Index.cfm?pageid=3292 [Accessed 5 June 2010].

Souter, . N. & Billout, G., 2007. *Breakthrough thinking, using creativity to solve problems.* Lewes, UK: ILEX.

Sternberg, R. J., 1994. *Thinking and Problem Solving: handbook of perception and cgnition.* London: Academic Press.

Swan, C., 2002. Action Research and the Practice of Design. *Design Issues,* Winter, 18(2), pp. 49–61.

Swann, W. B., 1987. Identity Negotiation: Where two roads meet, Personality and Dyadic. *Journal of Personality and Social Psychology,* 53(6), pp. 1038–1051.

Thomson, A., 2009. *Critical Reasoning: a practical introduction.* New York : Routledge.

Design of Information-Intensive Systems Involving Cognitive Aspects: An Emerging Opportunity for Transdisciplinary Cooperation

Regine W. Vroom and Wilhelm Frederik van der Vegte

Abstract With the rise of smart systems, ubiquitous computing and cyber-physical systems, information-intensiveness of products increases and users become challenged—possibly even overloaded—with expanding options and possible interactions. The number of possible variations of user-operation sequences can rapidly escalate and for designers it becomes difficult to foresee all possible outcomes, which might include unacceptable performance, failure, and even fatalities. With the objective to reduce the risk of unwanted cognitive effects and to realize a more symbiotic relationship between users and systems, we show how two model-based theories from cognitive science, i.e., cognitive architectures and mental models, can be deployed in the design of these systems. We argue that the deployment of such models requires a transdisciplinary approach in which designers intensively cooperate with cognitive scientists and end users.

Keywords Cognitive engineering · Information-intensive systems · Mental models · Cognitive architectures · Transdisciplinary cooperation

1 Introduction

Information-intensive systems (IISs) have been defined as systems where the use and production of information is either a major function or a major component of the control of the process. Such a system usually has as its components hardware and human beings, using software and procedures, respectively (Yamamoto et al. 1982). These systems have been part of our everyday lives for decades, as they include telecommunications, the electric grid, banking and financial services, manufacturing, surface transportation, petroleum delivery and emergency services (Jones

R. W. Vroom
Faculty of Mechanical, Maritime and Materials Engineering, Delft University of Technology, Delft, The Netherlands
e-mail: R.W.Vroom@tudelft.nl

W. F. van der Vegte (✉)
Faculty of Industrial Design Engineering, Delft University of Technology, Delft, The Netherlands
e-mail: w.f.vandervegte@tudelft.nl

L. Blessing et al. (eds.), *The Future of Transdisciplinary Design*,
https://doi.org/10.1007/978-3-319-06382-9_4

2000). IISs increasingly take the form of their current manifestations known as the internet of things (Horváth and Gerritsen (2012), meta-products (Huisman et al. 2011) and cyber-physical systems Lee 2006). They are often deployed in product-service systems (Boehm and Thomas 2013). As a consequence of their complexity, IIS development involves several aspects of product design (electronics, software, interface, communication, mechanics, robotics, industrial design, etc.), but typically also task design, organisation design and service design. The disciplines involved in designing the first generations of IISs were, among others, information and communication technology (ICT), industrial design engineering and mechanical engineering. Integrating cognitive psychology issues will be a key challenge in developing the next generations of IIS.

For various types of IISs it has already been argued that they require a transdisciplinary approach (Huisman et al. 2011), or inter-, multi- and transdisciplinary at the same time (Horváth and Gerritsen 2012). In the next section, we will present our interpretation of what 'transdisciplinary' is, considering the supporting literature and the context of IISs, as well as the distinction between trans-, multi-, interdisciplinary etc. Then, in Sect. 3 we elaborate on cognitive aspects of interacting with and designing IISs. We have identified these as an opportunity to set out directions for transdisciplinary cooperation, two of which are further elaborated in Sects. 4 and 5, respectively. The chapter wraps up with the discussion and conclusions in Sect. 6.

2 Transdisciplinarity in the Context of Developing IISs

According to Horváth and Gerritsen who discuss cyber-physical systems (CPSs) in Horváth and Gerritsen (2012), *interdisciplinarity* involves two knowledge domains (for CPSs: the cyber and physical domains), *multidisciplinarity* involves more than two knowledge domains (e.g. biology, engineering and computer science), and *transdisciplinarity* extends the knowledge from the various domains towards implementation and application, for instance by providing architectures and technologies to realize the artefacts and services within the CPS. This CPS-specific interpretation of transdisciplinarity seems to be in agreement with Pohl's description of transdisciplinary *research*, which he says 'is not only about producing knowledge but it is also problem- and solution-oriented, and the research results are translated into usable products' (Pohl 2000).

Generalizing these statements regarding what the various 'disciplinarities' mean in terms of cooperation between disciplines, we have concluded that they describe different types of professional activities in two dimensions, one dimension being that of the different domains (such as healthcare, agriculture, education) and the other being the conventional knowledge value chain, *research* → *design & development* → *application*, although other chains have also been suggested (Max-Neef 2005). Regarding cooperation, Wickson et al. (2006) signify that the intensity of the work requires mutual interactions between stakeholders over the concerned dimension(s), rather than that they access prepared knowledge from each other's domain—e.g., from books.

In addition to consulting literature about crossing disciplinary borders, we can learn from literature concerning the crossing of geological borders by companies, where the analogous terms inter-, multi- and transnational are commonly used. In that context, the set of definitions by Bartlett (1986) is often cited. In our context, the *differences in handling knowledge* that he has identified are the most relevant: *international companies* operate in multiple countries with knowledge developed at a central location and transferred to overseas units, *multinationals* develop and retain knowledge within each unit across multiple countries, and *transnationals* develop and share knowledge worldwide.

Based on the above assertions we have defined the different 'disciplinarities' as follows: ***Monodisciplinarity*** (mono- from Greek μόνος: *alone, only*) is confined to one domain, at one level of the knowledge value chain. An example is a project in which domestic-appliance engineers and designers are developing a coffee maker on their own. Knowledge from science or from users is purely used in an input-only fashion, e.g., from textbooks or available user surveys. ***Intradisciplinarity*** (intra- from Latin *within*) involves collaboration at multiple levels within the same domain. As an example, domestic-appliance engineers and designers are developing a coffee maker in close collaboration with end users and/or food scientists. ***Interdisciplinarity*** (inter- from Latin *among, between*) and ***multidisciplinarity*** (multi- from Latin *many*) are based on collaboration between different domains at one level. In interdisciplinarity one domain acts as a core domain, coordinating the other domains that supply contributions from their fields. Multidisciplinary cooperation is decentralized in that each involved discipline manages its own activities, based on cooperative central coordination. Consider for example, a project in which domestic-appliance engineers and designers are developing a pill dispenser for consumers (who may be also patients) in cooperation with a medical company. In the case of interdisciplinary design, the medical company acts as the principal and the dispenser has to conform to a given design of the pills or their packaging, whereas in the case of multidisciplinary design both parties deliberate over the requirements and specifications for both the dispenser and the pills. Figure 1 illustrates how we have interpreted these first four 'disciplinarities', taking the profession of engineering design as a starting point for reasoning.

Transdisciplinarity (trans- from Latin: *across*) implies cooperation at multiple, or even *all* (Max-Neef 2005), levels in two or more value chains. This is shown in Fig. 2, where cooperation should span at least one of the diagonal arrows or two of the horizontal arrows. In addition transdisciplinary (TD) activities may be inter-/multi-/intradisciplinary at the same time. As TD *research* has been defined as research involving translation of research findings into solutions (i.e. design), we can reason that TD *design* strongly depends on cooperation with researchers and/or end users. This suggests that there is no distinction between TD design and TD research and that it may be better to speak of TD projects or activities. In addition to the *relational* aspect of cooperation, we also consider the level of *maturity* of the connections between distinct disciplines relevant in characterising transdisciplinarity. Typical TD

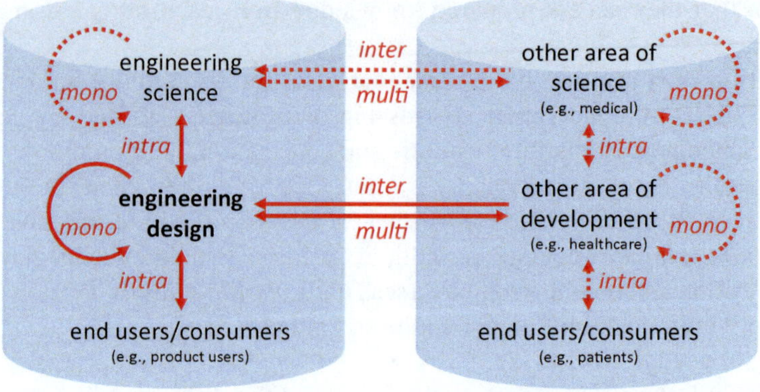

Fig. 1 Illustration of the meanings of *monodisciplinary, interdisciplinary, multidisciplinary* and *intradisciplinary*, reasoning from the field of engineering design

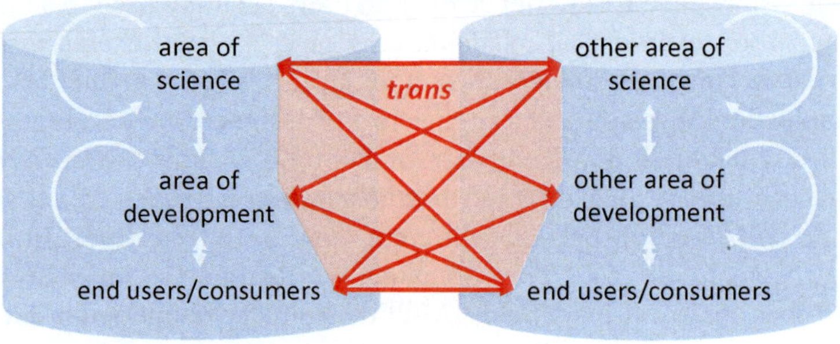

Fig. 2 Various forms of transdisciplinarity (elaborating on Fig. 1)

projects are *pioneering* efforts to connect domains. If relations become established over time, a new 'vertical' discipline is formed, and projects are no longer TD.

It has to be noted that the above definitions give a simplified view on the subject matter. Firstly, disciplines can be considered at various levels of abstraction. Therefore, the scope of 'disciplinarities' also depends on the observer's level of abstraction. Engineering for instance has many subdomains. At a lower level of abstraction a design project involving mechanical and civil engineering can be considered interdisciplinary, whereas it would be mono-disciplinary according to Fig. 1. Secondly, more layers can be distinguished in the knowledge value chain than the figures show. For instance, between *engineering design* and *end users* one could think of *manufacturing*, *distribution*, etc. Likewise, in the medical chain on the right hand side a layer *doctors* between *development* and *end users/consumers*, or even parallel to *development* can be added. And, as another contribution in this book shows, in urban infrastructure design, also the activities performed by stakeholders such as regulatory bodies and authorities can be recognized as parts of the value chain (Leblanc 2014). These additional layers could all be involved in interdisciplinary and TD projects. The number of layers in a chain can differ and it is not always obvious which ones are on the same level. As a third and final remark, the lowest level in the chain does not always show clearly distinct domains: a consumer who buys a coffee maker at one time can be a patient at some other time.

3 Cognitive Aspects and Issues of IISs

This section elaborates on one typical characteristic of IISs that we think requires a TD approach, namely that, in the way they are designed and the ways in which they function, IISs address issues of *human cognition* as well as *artificial cognition*. IISs will increase the level of communication and knowledge conversion technologies built into consumer products and systems. On the one hand IISs can take over particular cognitive tasks from users. Therefore, IIS designers will have to allocate cognitive tasks between user and IIS, and to design outputs of IISs to be relevant for users.

On the other hand, IISs are part of the information society that produces ever-increasing amounts of available information, both valuable and useless. It means that besides *reducing* cognitive task loads, IISs may also confront users with *increased* amounts of information. The increase may negatively influence use comfort, and cause perceptual and/or cognitive overload in demanding situations. In addition, it is expected that, since they offer functionalities that cannot be realized with conventional technology, IISs will increasingly be deployed in safety–critical situations (Karnouskos 2011). From conventional safety–critical systems, such as nuclear plants, it is known that their evaluation involves identification of rarely occurring scenarios, e.g., once in 1,000 years (Beckjord et al. 1993). In many circumstances where IIS will play an increasingly important role, such as car driving, air traffic and medical care (Baheti and Gill 2011; Lee and Sokolsky 2010; Work et al. 2008), we also have to consider infrequent scenarios (e.g. likelihood once in 500 years per driver/pilot/physician) in risk assessment. This is only possible by comprehensively understanding human cognitive behaviours under varying situations including emergency or stressful scenarios (Poovendran 2010).

We believe that this increased understanding will eventually enable designers to realize cognitive symbiosis between IISs and humans, from which not only safety–critical IIS but also IIS supporting everyday life, will benefit—for instance by increasing user comfort and satisfaction. Achieving symbiosis is expected to be more important for IIS intensively interacting with humans, and arguably less for autonomous IIS acting without any human intervention. The desired symbiosis requires knowledge from cognitive science, and development of design tools in cooperation with cognitive scientists. In addition it is likely to also require involvement of end users, both in research and in design activities. It is expected that, because the cooperation spans across the knowledge value chain *and* various domains, the resulting TD cooperation will pose several challenges to the stakeholders involved. It means we have to deal with (i) different jargons used by experts of electronics, software, interface, communication, mechanics, robotics, industrial design, (subdomains of) cognitive science and other domains, (ii) different work attitudes of the people involved (e.g., synthesis-oriented and result-driven vs. analytical and curiosity-driven), and (iii) different ways of evidencing and validating the outcomes of the work (e.g., calculations vs. empirical testing).

In this chapter we will briefly discuss two directions of research in which we aim to study how knowledge from the cognitive sciences can be used in ISS development processes, addressing the above issues at two levels. The first one, introduced in Sect. 4, aims to use *cognitive simulations* in order to identify potential bottlenecks for human information processing as well as options to resolve them. In this context it is assumed that the IIS and its use scenarios have been worked out to such an extent that they can be modelled and simulated. The second direction of research, introduced in Sect. 5, addresses the issues at a higher level in order to support the early stages of designing IISs. The aim is to gain operational knowledge on *mental models* that can be used to design better informing systems. People use cognitive representations in order to characterize, understand, reason and predict the surrounding world. A class of these representations are called mental models (MMs). Designers of informing systems need predictive power on the knowledge and reasoning patterns of potential users of their systems. The concept of MMs, is expected to provide the basis for the minimal required understanding of the human reasoning.

As authors of this contribution, we are operating at the science level in Figs. 1 and 2 in both of these initiatives. Our interest is to investigate new ways of supporting designers. Our work combined with contributions by designers who implement the results represents the 'design engineering' side of the projects. In Sects. 4 and 5 we will mostly focus on explaining the cognitive-science involvement.

4 Simulating Cognitive Loads and Processing Times

The first research direction concerns a plan conceptualized together with cognitive scientists to develop an approach for co-simulating human mental processes and models of products and systems. The goal is to evaluate IISs during development, in order to identify bottlenecks that need to be resolved by adapting the design – i.e., the design of the system, the design of human tasks or the related service design.

We propose to test IISs without humans in the loop by using a *cognitive architecture* (CA) as a model of human information processing and decision-making. One project concerns simulation of centralised pound-lock control (PLC) rooms, to be operationalised by our government agency of public works from 2014 onwards. The second category of IISs that we consider for conceptualization and study is emergency response systems (ERSs) in buildings. ERSs currently involve several systems and devices, some of which operate connectedly to facilitate a variety of situations, including fire detection, medical assistance, communication with firefighters/paramedics/police and managing evacuations. Although some of today's systems and products are technologically quite complex, we see considerable potential in further integrating and enhancing them based on cyber-physical technologies, e.g., advanced detection based on sensor networks, intelligent proactive assistance and ad-hoc communication networks. Systems with some of these technologies have already been prototyped, but so far mostly focusing on victim monitoring by medics at large-scale disaster sites (e.g. Gao et al. 2008). We expect additional challenges when dealing with, for instance, non-expert volunteers, evacuation of buildings and isolated but more frequently occurring incidents and drills.

Both the PLC system and ERSs nicely illustrate the potential of our approach for IIS designers because they are safety–critical, and the IIS acts in close cooperation with human operators who are still in charge of important decisions. The operators have a high responsibility to act according to protocols that involve taking into account many different factors. For PLC this includes dealing with various lay-outs of locks, types of boats and skippers, weather circumstances, etc. In addition, most locks have multiple chambers, in connection with which the newly introduced procedure of 'zipper-wise operation' increases the operators' multitasking load. In exceptional cases, cognitive processing errors by operators may lead to severe accidents or even disasters (colliding ships, flooding). Likewise, emergency response workers have to make split-second decisions for instance about which actions they can perform themselves and which ones are best left to fire-fighters and paramedics – in a wide range of situations including heart failure and escalation of a fire, which can obviously present themselves as matters of life and death.

Due to the limitations of real-time simulation, it is impossible to use an interactive simulator for testing all combinations of factors and sequences of occurrence. However, by combining simplified system models with ACT-R (adaptive control of thought–rational)—a CA that has proven to produce accurate, scientifically validated simulations of the relevant phenomena, i.e., multitasking, cognitive overload, distraction, fatigue, memorising, etc. (Salvucci et al. 2009)—we expect to run the simulations much faster than real-time, so that even rare critical situations can be revealed (Vegte and Moes 2012).

CAs are blueprints of cognition based on findings from brain science. Figure 3 shows ACT-R's modules and the identified corresponding areas in the human brain (Anderson et al. 2004). The *external world* block corresponds to everything outside the human. In our pond lock example it would comprise the operation interface, the

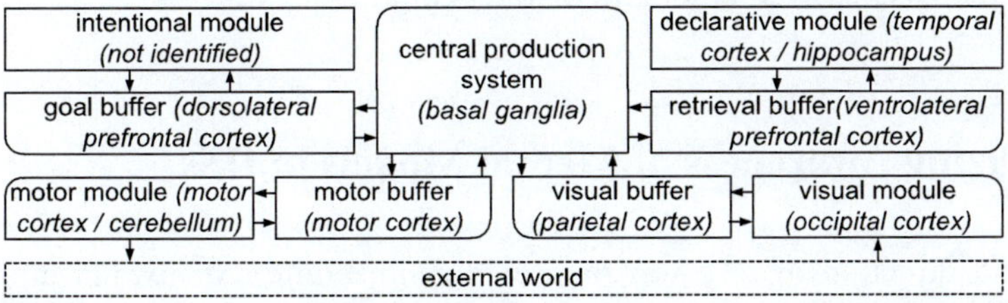

Fig. 3 Modules of ACT-R and corresponding cortical regions (*in italics*). Adapted from Anderson et al. (2004)

locks themselves with related constructions (bridges, traffic lights, etc.) and, based on available statistics, the traffic and the weather. The connection between ACT-R and the external world is established through the motor module (human output through control of limbs) and visual module (human input through visual perception). In case of aural input, an aural module is included as well.

Simulation models in ACT-R are always custom-built for a specific case. Each module is 'filled' with routines programmed in LISP describing information-processing behaviour related to specific subtasks (Patterson et al. 2013). For common subtasks, LISP routines are readily available; for others, laboratory studies with human subjects have to be conducted to collect data for new routines. The overall task of the human, e.g., the protocol for operating pond locks, is written as a LISP routine for the intentional module. Laboratory experiments and programming of routines are activities that require expert knowledge about cognitive information processing. Therefore, in its current form ACT-R is mainly used by cognitive scientists and it is not an off-the-shelf simulation tool for designers. Consequently, its embedding in design calls for a TD approach in cooperation with cognitive scientists.

In this cooperation there is also a strong aspect of *pioneering*. Although application of CAs has already become more practical—evolving from puzzle-solving i.e., pure brain exercises with 'disembodied' CAs lacking visual and motor modules Anderson et al. 1997), through interactions with software via mouse, keyboard and monitor (Byrne 2005), to specific tasks in aviation (Byrne and Kirlik 2005) and car driving (Salvucci 2006)—they have not yet been applied in interaction with complex multi-faceted external world models, despite obvious potential benefits. A possible explanation is that, on the one hand designers are not aware that simulation of mental processes is actually possible, and that on the other hand cognitive scientists come from a research tradition of controlled experiments that benefit from simple external worlds. To promote pioneering in TD projects, we therefore have to facilitate designers in utilising research efforts that can contribute to their work, and find ways to make researchers benefit from practical applications. In the case of PLC simulations, this might involve developing validation methods for outcomes like 'once in 1,000 years, a cognitive operator error will cause flooding', which cannot be straightforwardly verified in a controlled experiment.

The investigation of mental models in the next section involves cooperation with cognitive scientists as well. However, cognitive architectures and mental models require different investigative approaches (laboratory measurements vs. interviews), and the scientists involved belong to distinct communities.

5 Realizing Awareness of Mental Models in IISs

The second direction of research focuses on *informing systems* and aims to find novel means to inform users and to find new symbiotic relations between human and systems, based on which designers can be supported in the early stages of IIS development. In cognitive psychology the internal representation that people hold

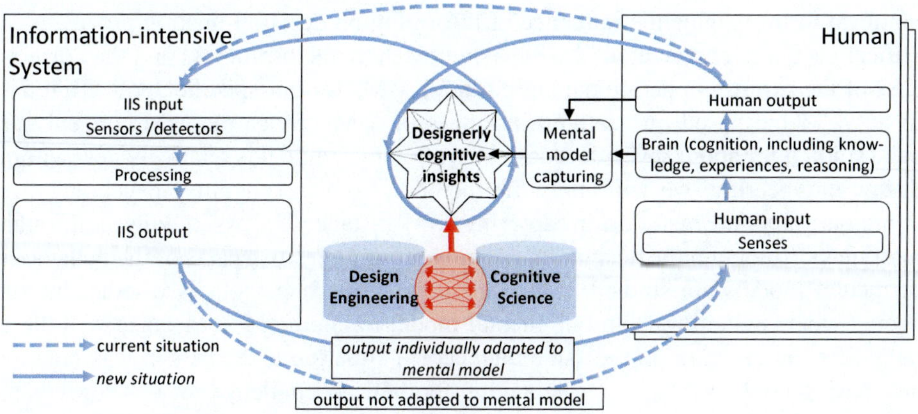

Fig. 4 Cognitive insights to influence the adaptability of IISs

of an external reality that allows them to explain, interact, and predict that reality is called a mental model (MM). MMs have been identified as a basis of human reasoning (Johnson-Laird 2010). This makes the phenomenon an interesting starting point to consider in designing the human interface of systems. The purpose of this project is to gain a better understanding in the manner in which MMs influence our interaction with IISs, and to provide guidelines for designers based on these insights. The project will therefore produce a predictive theory and additionally formulate its affordances for the design process. In Fig. 4 it is shown that human output can be directly detected by the system. Currently this is detected through for example motion detection, id-tags or smartphone detection. In the future desired situation the cognitive implications in the human output will be interpreted with the obtained designerly cognitive insights enabling the system to adapt its output to the cognitive capabilities of an individual user in a specific situation. It may give specifically the information that will help to take a right decision to react.

For this project, new insights are needed about the operation of MMs, as well as on how, for our specific design objective, the real-life operationalization of MMs is influenced by informing. We have assumed that the highest need for adapting the level and content of the provided information will be in critical situations that cannot be anticipated straightforwardly. Therefore, the objective of the first phase is to address this problem by deriving a definition of MMs which, in contrast to already existing definitions, will be tailored to critical events. Since the definition has to be both meaningful in our specific context of designing highly adaptive IISs and correct regarding its psychological fundamentals, the disciplines of design engineering and cognitive psychology have to be fused.

The expertise from cognitive psychology was initially adopted from selected relevant scientific papers: 125 published descriptions of MMs have been decomposed to a set of attributes, and each attribute has been assessed to see if it was associated with critical events. This exploration provided a large number of attributes for a new MM definition. Based on the top-rated attributes, a definition was synthesized as a starting platform to investigate the influence of informing on decision-making processes in critical events (Deurzen et al. 2013). As a next step, the usefulness and the correctness of the resulting operational definition of MMs for our specific application has to be validated based on captured instances of MMs. Since the MM concept has been studied for about seven decades in psychology, while it is relatively new within design engineering, we will apply the methods from the cognitive psychology. A commonly applied method for capturing MM instances in psychology is through interviews. To obtain the designerly cognitive insights, two aspects of the behaviour of mental models are studied. A first element of the behaviour is whether inertia occurs when switching from one mental model to another. For instance will there be a different reaction on the same unexpected situation if the person was reading an exciting book as when he was playing football? A challenge for this study is to cope with the irreversibility of perceptions that occur even in experimental set-ups. A second contribution to the designerly insights will be the exploration and development of a method to identify inadequacies in a person's knowledge and experience. In cognitive psychology it is commonly accepted that mental models are inaccurate and incomplete (Sonnentag 1998). Gained insights in identifying the gaps and faults in a MM will indicate ways to "repair or improve the MM" which means to better inform people. Subsequently, a study to effectively address the insufficiencies in a mental model will constitute the bridge towards guidelines for designers to develop IISs with adaptive capabilities on the user's cognition. These aimed guidelines for addressing the gaps and faults in a MM will include the contents, the senses to address and the effect of the amplitude of the message, being e.g. the volume of aural information or the pressure level of haptic information.

Hence predictive power will be inferred from the captured instances by exposing them to selected events and monitoring the effects on human reasoning and behaviour. These data will be analysed to find cause-effect relationships. From these discoveries, theories will be derived describing the behaviour of MMs for specific events. Both for validation and evaluation of the operational construct of an MM for our specific objectives, and for the elaboration towards predictive functionality based on new theoretical insights, we will reach a point where we either have to become an expert in the field or find close cooperation with cognitive psychologists to fuse the knowledge and methods. To verify the obtained results and to elaborate on the new insights, the expertise of a cognitive psychologist is expected to add more value than can be achieved through solely reading and applying published results.

6 Discussion and Conclusions

In this chapter we discuss transdisciplinary cooperation between design and research in the context of IISs. We started out from setting transdisciplinarity apart from (in particular) multidisciplinarity and interdisciplinarity, which may have led to a somewhat stricter interpretation of transdisciplinarity than has been proposed in other contributions to this book. However, our definition can be used in accordance with most of the other definitions and descriptions that have been brought forward during the workshop and in this book, such as the often-cited assertions that transdisciplinarity should integrate beyond the boundaries of the contributing disciplines and that it is holistic [cf., Fernandez-Orviz (2014), Gericke (2014)]. The definitions appear to agree that transdisciplinarity involves more mutual commitment between disciplines than multidisciplinarity or interdisciplinarity.

We are currently exploring two directions of transdisciplinary design/research in the interfacing area between engineering and cognitive science. The practical application potential of knowledge from cognitive sciences to design engineering problems is still largely unexplored. The first presented research direction aims to deploy cognitive architectures (CAs) in evaluating designs of safety–critical IISs and minimize harmful cognitive effects, and in the second one we aim to consider the concept of mental models (MMs) as a carrier to harmonize the information exchange with systems to the user's expectations and reasoning patterns. We expect that adoption of such approaches in IIS design will eventually result in optimally symbiotic relations between humans and the increasingly complex systems around them. Regarding the two, seemingly closely related, directions of research and their transdisciplinarity, we would like to conclude with two observations. One concerns the recognition of 'cognitive science' as one monolithic discipline, the other concerns the recognition of 'experts' from another field in general.

An obvious future step in our work would be to expand the transdisciplinary scope and combine MMs and CAs in one design-support approach. A possible challenge in such a cooperation is that it may necessitate cooperation between disjunct research communities within cognitive science, who even might represent diametrically different viewpoints on how the human brain works and how it should be investigated.

Regarding the decision to involve *experts* from other disciplines, the need arises to reflect on the distinction between experts and non-experts. Alexander (2003) states that characterizations of expertise were traditionally based on sharp contrasts between experts and neophytes, but that in fact, subtle and significant transformations occur between those extremes. Ahmed et al. (2005) and Sonnentag (1998) express the distinction between non-expert and expert in years of experience. Based on interviews, Ahmed et al. found that, in the field of engineering design, someone is considered an expert after 5–15 years of relevant experience, while Sonnentag argued that expertise in software engineering requires at least ten years. Apparently, there is no sharp definition of 'expert' that can be used to decide whether a partner contributing knowledge from another discipline is an expert and consequently makes a project a TD project.

References

Ahmed S, Hacker P, Wallace K (2005) The role of knowledge and experience in engineering design. In: Proceedings ICED. Melbourne

Alexander PA (2003) The development of expertise: The journey from acclimation to proficiency. Educational Researcher 32:10–14

Anderson JR, Bothell D, Byrne MD et al. (2004) An integrated theory of the mind. Psychological Review 111:1036–1060

Anderson JR, Matessa M, Lebiere C (1997) ACT-R: A theory of higher level cognition and its relation to visual attention. Human-computer interaction 12:439–462

Baheti R, Gill H (2011) Cyber-physical systems. In: The Impact of Control Technology. IEEE Control Systems Society, p 161–166

Bartlett CA (1986) Building and managing the transnational: The new organizational challenge. In: Porter ME (ed) Competition in global industries. p 367–401

Beckjord ES, Cunningham MA, Murphy JA (1993) Probabilistic safety assessment development in the United States 1972–1990. Reliability Engineering & System Safety 39:159–170

Boehm M, Thomas O (2013) Looking beyond the rim of one's teacup: A multidisciplinary literature review of Product-Service Systems in Information Systems, Business Management, and Engineering & Design. Journal of Cleaner Production 51:245–260

Byrne MD (2005) Cognitive architectures in HCI: Present work and future directions. In: Proceedings Int. Conf. on Human Computer Interaction. Las Vegas

Byrne MD, Kirlik A (2005) Using computational cognitive modeling to diagnose possible sources of aviation error. Int. Journal of Aviation Psychology 15:135–155

Fernandez-Orviz A (2014) Ineffective Collaboration in Multi-disciplinary Teams. In: This book. Springer

Gao T, Pesto C, Selavo L et al. (2008) Wireless medical sensor networks in emergency response: Implementation and pilot results. In: Proceedings Conf. on Technologies for Homeland Security. IEEE, p 187–192

Gericke K, Adolphy S, Qureshi A et al. (2014) Opening up Design Methodology. In: This book. Springer

Horváth I, Gerritsen B (2012) Cyber-physical systems: Concepts, technologies and implementation principles. In: Proceedings TMCE. Karlsruhe, p 19–36

Huisman M, Hazenberg W, Cordoba Rubino S (2011) Meta-products: meaningful design for our connected world. BIS Publishers, Amsterdam

Johnson-Laird PN (2010) Mental models and human reasoning. Proceedings of the National Academy of Sciences 107:18243–18250

Jones A (2000) The Challenge of Building Survivable Information-Intensive Systems. Computer: 39–43

Karnouskos S (2011) Cyber-physical systems in the smart grid. In: Proceedings Int. Conf. on Industrial Informatics. IEEE, p 20–23

Leblanc T (2014) Design as a common denominator. In: This book. Springer

Lee EA (2006) Cyber-physical systems: are computing foundations adequate? In: Proceedings Workshop On Cyber-Physical Systems: Research Motivation, Techniques and Roadmap. NSF, Austin, p 1–9

Lee I, Sokolsky O (2010) Medical cyber physical systems. In: Proceedings Design Automation Conference. ACM, Austin, p 743–748

Max-Neef MA (2005) Foundations of transdisciplinarity. Ecological economics 53:5–16

Patterson RE, Fournier LR, Williams L et al. (2013) System Dynamics Modeling of Sensory-Driven Decision Priming. Journal of Cognitive Engineering and Decision Making 7:3–25

Pohl C (2000) Inter-and transdisciplinary research methods. In: Transdisciplinarity: Joint problem-solving among science, technology and society. Workbook I: Dialogue Sessions and Idea Market. p 18–19

Poovendran R (2010) Cyber–Physical Systems: Close Encounters Between Two Parallel Worlds. Proceedings of the IEEE 98:1363–1366

Salvucci DD (2006) Modeling driver behavior in a cognitive architecture. Human Factors 48:362

Salvucci DD, Taatgen NA, Borst JP (2009) Toward a unified theory of the multitasking continuum: From concurrent performance to task switching, interruption, and resumption. In: Proceedings SIGCHI. ACM, Boston, p 1819–1828

Sonnentag S (1998) Expertise in professional software design: A process study. Journal of Applied Psychology 83:703

Van Der Vegte WF, Moes NC (2012) Towards Improved User-Product Testing With Cognitively Enhanced Scenarios. In: Proceedings CIE. ASME, Chicago

Van Deurzen K, Horvath I, Vroom R (2013) Defining the phenomenon of mental models for critical events. In: Proceedings CIE. ASME, Portland

Wickson F, Carew AL, Russell A (2006) Transdisciplinary research: characteristics, quandaries and quality. Futures 38:1046–1059

Work D, Bayen A, Jacobson Q (2008) Automotive cyber-physical systems in the context of human mobility. In: Proceedings Nat. Workshop on high-confidence automotive cyber-physical systems. Troy

Yamamoto Y, Morris RV, Hartsough C et al. (1982) The role of requirements analysis in the system life cycle. In: Proceedings ACM National Computer Conference. ACM, p 381–387

Visually Augmented Analysis of Socio-Technical Networks in Engineering Systems Design Research

M. Štorga, T. Stanković, P. Cash, and T. C. McAloone

Abstract In characterizing systems behaviour, complex-systems scientists use tools from a variety of disciplines, including nonlinear dynamics, information theory, computation theory, evolutionary biology and social network analysis, among others. All of these topics have been studied for some time, but only fairly recently has the study of networks in general become a major topic of research in complex engineering systems. The research reported in this paper is discussing how the visually augmented analysis of complex socio-networks (networks of people and technology engaged in a product/service-system (PSS) life cycle) may be applied in engineering design research. Network thinking of the kind described in this paper could be fundamental for developing new and effective techniques for solving the problems in the engineering design research related to the interpretation of the huge amount of data captured during experiments and observations that are more and more used as a main research method. Case studies that are presented illustrate also the significance of the network based research approach in providing insight into ways of improving the design process for complex engineering systems.

Keywords Socio-technical networks · Visualisation · OrganicViz · Traceability · Engineering design research

M. Štorga (✉)
Faculty of Mechanical Engineering and Naval Architecture, University of Zagreb, Zagreb, Croatia
e-mail: mario.storga@fsb.hr

T. Stanković
Engineering Design and Computing Laboratory, ETH Zürich, Zürich, Switzerland

P. Cash
Department of Technology, Management and Economics, TU Denmark, Lyngby, Denmark

T. C. McAloone
Department of Mechanical Engineering, TU Denmark, Lyngby, Denmark

L. Blessing et al. (eds.), *The Future of Transdisciplinary Design*,
https://doi.org/10.1007/978-3-319-06382-9_5

1 Introduction

In the recent decade, understanding the structure and function of complex systems has become the foundation for explaining many different real-world complex natural, social and technological phenomena. Often-cited examples of complex systems are, social networks of acquaintance or other connections between individuals, organizational networks and networks of business relations between companies, the World Wide Web, neural networks, metabolic networks, food webs, distribution networks, networks of citations between research papers. Although there is no generally accepted definition of "complex system", informally, a complex system is a large network of relatively simple components with no central control, in which emergent complex behaviour is exhibited (Mitchell 2006). The complexity of the system's global behaviour is typically characterized in terms of the patterns it forms, the information processing that it accomplishes, and the degree to which these pattern formation and information processing are adaptive for the system - that is, increase its success in some evolutionary or competitive context. In characterizing behaviour, complex-systems scientists use tools from a variety of disciplines, including nonlinear dynamics, information theory, computation theory, behavioural psychology, evolutionary biology and social network analysis, among others. All of these topics have been studied for some time, but only fairly recently has the study of networks in general become a major topic of research in complex engineering systems. In addition, recent technological advances produce a lot of data and have led to the establishment of large and complex network models across various domains. The engineering systems design research is not an exception. Therefore, the main motivation for this paper is to examine and illustrate how the visually augmented analysis of complex socio-networks (Mostashari 2010) (networks of people and technology engaged in a product/service-system (PSS) life cycle) may be applied in engineering design research and discusses the significance of the network based research approach in providing insight into ways of improving the design process for complex engineering systems.

2 Background/Related Works

2.1 Complex Networks

A network is a set of items, which we will refer to as *nodes*, with connections between them, called *edges*. The mathematical study of networks arose from graph theory, which began as early as the eighteenth century when Euler established the famous "Bridges of Königsberg" problem. However, until recently, mathematical graph theory did not have a large impact on the study of real-world networks, since the latter's properties were quite different from those of random graphs (Newman 2003). Networks have also been studied extensively in the social sciences. Typical social network studies address issues of centrality (which individuals are best connected to others or have most influence) and connectivity (whether and how individuals are connected to one another through the network). In the recent decade we have witnessed a substantial new movement in network research, with the focus shifting away from the analysis of single small graphs and the properties of individual nodes or edges within such graphs to consideration of large-scale statistical properties of graphs (Newman 2003). In this period increasing numbers of applied mathematicians, physicists, social scientists and computer scientists have joined together in developing a general theory of networks. Techniques from statistical physics have been successfully applied to the analysis of complex networks and have uncovered surprising statistical structural properties that have also been shown to have a major effect on their functionality, dynamics, robustness, and fragility (Braha and Bar-Yam 2007). Among the reasons for this are fast computers and new algorithms being developed, which make it possible to study real networks empirically, and the increased attention this field is getting from physicists, looking to other fields to apply their powerful analysis techniques. Here are some of the questions that network scientists are trying to address:

- What topological metrics can be used to characterize properties of networks?
- What properties do different sets of real-world networks share, and why? How did these properties come about?
- How do we design efficient algorithms to determine these properties?
- How do these properties affect the dynamics of information (or disease, or other communications) spreading on such networks, and the resilience of such networks to noise, component failures, or targeted attacks?
- Given a network with certain properties, what are the best ways to search for particular nodes in this network?

It is reasonable to believe that answering these questions could have a large impact, not only on our understanding of many natural and social systems, but also on our ability to effectively use complex networks in engineering systems design research. Potential applications include better engineering information search, controlling the spread of innovation, understanding the evolution of the knowledge structure in development projects, management of the cognitive organizations, understanding the PSS life cycle, or predicting the potential damage resulting from interaction of human actions and engineering systems in socio-technical networks.

2.2 *Complex Networks in Engineering Systems Design Research*

Planning techniques and analytical models that view the engineering systems design process as a network of interacting components have been proposed in literature (Braha and Maimon 1998; Yassine and Braha 2003; Klein et al. 2006; Mihm et al.

2003). However, those research projects have not addressed the large-scale statistical properties of real world product development based networks. Braha and Bar-Yam (2007) have shown that product development task networks have properties (sparseness, small-world, scaling regimes) similar to those of other biological, social, and technological networks. They discover distinctive asymmetry between the distributions of incoming and outgoing information flows (links) in product development networks, which has implications for their functionality, sensitivity, and robustness (error tolerance) properties. Network analysis was also applied to study task interactions in product development systems in the work of Colins et al. (2010). They concluded that network analysis offers a suite of metrics to analytically measure the information flow characteristics in complex product development environments, rather than relying on an individual or set of individuals to accurately understand the entire process. The core-periphery structure of complex product development processes has been further studied by Li and Xu (2011) from the perspectives of information flow and social networking. They model and analyse product development process as a graph which adjacency matrix is visualized using design structure matrix, and employ social network analysis to identify and explain the core-periphery structure in complex product development process.

Most of the described approaches were focused to the structural properties of the networks considered. However, in order to understand the behaviour of the system modelled as a network, it is necessary to explore the mechanisms by which the network is created over the time and how the structure and properties of the network evolve. For that purpose, dynamic network analysis methods and tools should be applied. It is known that good visualisation analysis reveals the hidden structure of the networks and amplifies human understanding, thus leading to new insights, new findings and possible prediction.

2.3 *Organic Visualisation of Complexity and Dynamics*

Organic information visualisation proposed by Fry (2008) offers an approach that employs a computer based scientific animation and simulation to conduct qualitative analysis of complex and dynamic network structures. Fry's approach assumes a metaphor towards living systems thus establishing a visualisation technique by mimicking organic properties and behaviour. The abstract entities employed for the organic visualisation in Fry's proposal are developing, evolving, interacting towards emergent behaviour, reproducing and eventually dying out, altogether creating a visualisation technique for understanding and interpretation of the complex network structure (Ogawa and Ma 2009).

Originally, the organic visualisation paradigm was applied to visually retell the story written in a book within a dynamic, computer supported, interactive environment. A rule based system, which processed the book's content linearly, was developed to create a three dimensional network with unique words being nodes connected together and grouped more closely, if found located next to each other within the text (Fry 2008). The more frequent appearing words where made to end up located on the 3D network's surface. Later, the same project was used to create a genome visualisation platform to visually inspect the occurrences of typical gene sequences (for more info please visit benfry.com). Another example of how organic visualisation is being employed to visualise a product development process can be found in the work of Ogawa and Ma (2009), where a history of developers' activity in software development is being visually analysed.

Organic systems are by definition open and information processing systems (Mitchell 2006), all of which exhibit responsiveness to stimuli and self-organization to maintain order whilst resource competing. Assuming such behavioural traits for non-organic real life systems, like engineering systems design processes, the idea of organic visualisation becomes compelling for visualizing the network structure evolution occurring when modelling these processes. This is further supported if we take into consideration the entire product/service-system (PSS) life cycle situated in a socio-technical context. The approach of Dong and Moere (2005) for example, utilises a three dimensional visualisation to study and understand dynamics of large-scale design team collaboration. In that work it was pointed out that conventional development project management representations of organisation charts and graphs about deliverables are not suitable for providing an understandable foundation from which team performance on a social level can be established.

3 OrganicViz Visual Analysis Tool

For the applications in industrial and manufacturing organisations Stanković et al. (2012) applied the paradigm of organic visualization (Fry 2008) to visually analyse dynamics of the information evolution. In order to model complex networks comprising of information evolution content and context a labelled directed multi graph was applied in OrganicViz tool (www.organicviz.org). For such model it turns out to be sufficient for describing the semantic network which is formed according to predefined domain ontology (Ahmed and Štorga 2009). In relation to the original organic visualisation (Fry 2008), the OrganicViz employs and extends the following features for the visualisation of information evolution in industrial and manufacturing domains.

- Structure—an aggregation of elements to form more complex structures; a directed labelled multi graph is applied for structuring; it is assumed that forming or clustering nodes into communities is dependent on the basis of ontology (Štorga et al. 2011b) that allows multiple relationships between the concepts to provide the domain of discourse.
- Growth—an increase in either scale or amount of structure; it is assumed that during the observed period of time the system that is modelled will increase in number of elements, objects and relations.

- Homeostasis—the maintenance of a balanced internal state; the labelled multi-graph's layout is maintained by "force field strength" which is calculated based on nodal connections within its neighbourhood. Fluid damping relaxation is applied to achieve smoothness in layout formation after stimuli have been applied.
- Responsiveness—reaction to stimuli and awareness of the environment; interaction is applied by various filters, direct interaction to graph entities or indirectly via graphical user interface controls.
- Adaptation—adjustments to survive in a changing environment; nodes cannot be deleted but can be hidden. The layout will always recalculate in respect to the pre-existing structure prescribed by the ontology and non-hidden nodes.
- Movement—the structure or its elements may be shifted in-plane, nodes can move as the result of the addition of new nodes to the existing structure to avoid overlap, or nodes can move as the result of new relational additions.
- Reproduction—the ability of entities to create others like themselves; new nodes that enter the visualized network over time are positioned depending on the ontology and pre-existing structure, thus being located in the vicinity of entities to which they are related.

4 Case Studies

The organic visualisation paradigm (Fry 2008) applied within the OrganicViz tool has been used in the context of engineering systems design research during several case studies. For each case study, the growth of the network was analysed in order to define the process as random or uniform (Newman 2003). As the second step, evolutionary dynamics were created by visualising network configuration changes over time. Next, the filtered viewpoints were applied on the evolving networks based on the context defined by specific case. Finally, a statistical examination of network formation was performed in order to assess the dynamic nature of network evolution. The specific cases studied are presented in more detail in the following sections.

4.1 Information Evolution Traceability

This case study examines the use of a semantic traceability record as an integrator of process and product related information. This brings together fragmented information across different information objects managed by various engineering support tools and is described as the main contribution of the TRENIN project (Štorga et al. 2011a). In order to validate the proposed visualization method and developed tools, the product development projects from two industrial partners in the automotive and energy sector were selected (Štorga et al. 2011b).

1. The development of new vehicle control systems for trams/trains. Vehicle control systems control all the electronics in electrical vehicles. As such, the control

unit is responsible for control, measuring, sequencing, protection, supervision and communication. For the purpose of validation several traceability record templates for the early development phase were developed. The traceability episodes were then executed in order to capture the information necessary for the visual analysis of the design project. Specifically, consolidation of the design requirements during development of a new system variant and the process of hardware and software subsystems testing after integration were considered.

2. The development of a new generation of vertically self adjusting head supports for car seats. Ergonomically designed head support encourages a relaxed sitting posture and can help with neck pain relief, lower back pain, tension and driving fatigue. The car seat head offers firm support and promotes a relaxed posture improving driving comfort as well as car seat safety. Several traceability record templates were created for the design project and visually analysed in order to trace information relating to the execution and implementation of safety guidelines and norms in components as well as the realisation of key product characteristic in order to ensure the quality of the solution accordingly to the requirements.

The complex heterogeneous networks based on the traceability record templates were created with the following elements. An example screenshot of the created network of information and information objects at one particular point in the traceability episode is shown in Fig. 1.

- Combination of the process and product oriented traceability elements—nodes.
- Traceability objects representing the information objects whose evolution was traced during the episode—nodes.

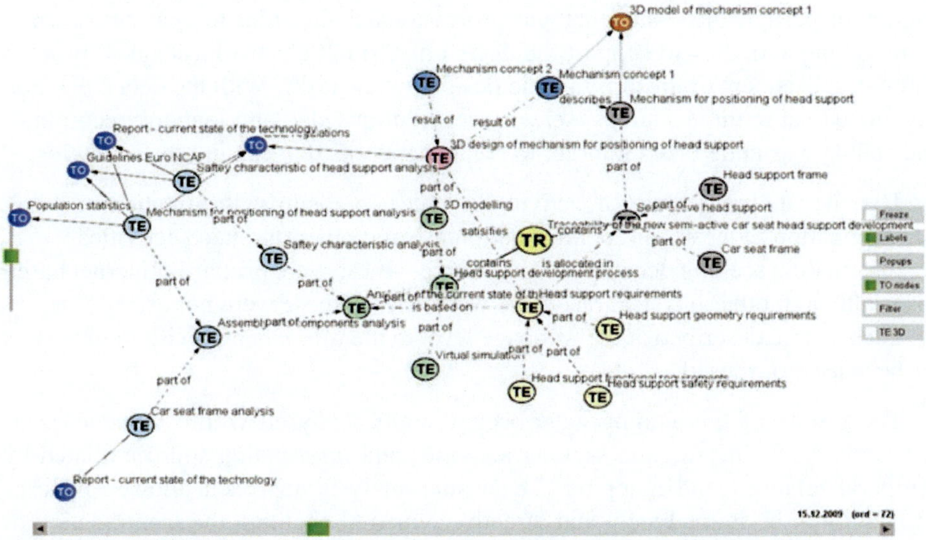

Fig. 1 Analysis of information and information objects evolution (Štorga et al. 2011a)

- Semantic relations of different types (influence, dependency, composition) describing the context of the information evolution during the traceability episode.

The validation objectives of the case study were focused to the demonstration of the complex network visualization approach's ability to support understanding and knowledge generation based on dynamic evolution of the recorded content. The experience of the users from the industrial partner confirmed that it was necessary to visually represent the dynamic of the traceability records for better understanding of the semantic structure of the information objects. Further, discussion with the participants of the study confirmed that traceability record evolution could be better understood and navigated, and that the information was easier to uncover, understand and maintain using the visual analysis of the interrelations over time.

For correct interpretation of the recorded traces it is essential to have a visible record structure together with the interfaces and procedures for searching and/or navigating through the structure. Comparison of the visual approach with textual reports based upon the same records confirmed that users were able to faster understand recorded content and reuse it in a proper way when they were supported by the dynamic network analysis tool (Pavković et al. 2013).

4.2 Complex Behaviour Patterns in Information Seeking Activities

A method for using information visualisation and statistical analysis to explore complex patterns in the activity of design practitioners during information seeking episodes were discussed by Cash et al. (2013). This experimental study used two groups of participants—students and professionals—in order to analyse information seeking activities to support the design of a small electro-mechanical product through subsequent brainstorming and design review tasks. With the data collection completed using three sources (screen, workstation video and logbook recording), the coding was undertaken in order to create networks that capture the following.

- The overall breakdown of activity in order to identify information seeking, information requests, direct information use and other unrelated activities.
- Information seeking decomposed by source—these were primarily Internet based as although other physical resources were offered these were not used.
- Each source described at the webpage level to identify when specific sources had been used or reused.

The results of this coding have been visually analysed within OrganicViz as the networks of information seeking activities and information sources related by temporal relations and mapping. The visual analysis approach allowed patterns of activity to be immediately and directly compared enabling the identification of common patterns or differences not immediately obvious using normal protocol analysis (common patterns of activity identified during the studies are given in Fig. 2).

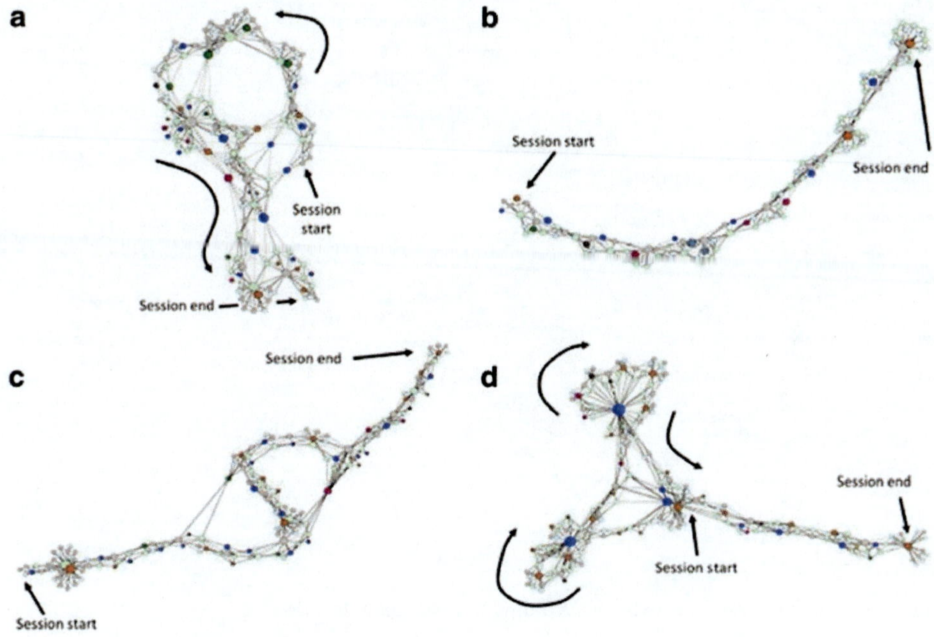

Fig. 2 Analysis of information seeking activities

Based on these patterns the results could be grouped into three main types.

1. Linear, with limited local iteration and a maximum of one larger scale iteration in the form of a loop e.g. Student 5 Fig. 3.
2. Mixed, with linear segments connecting multiple areas of large-scale iteration.
3. Complex, consisting entirely of iterative loops of various sizes and with no isolated linear segments.

In particular, by identifying different patterns of information seeking activity it was possible to group participants and focus analytical effort on determining the reasons for the differences, which are not apparent when simply considering either the total duration of each activity/source or how these changed over time individually.

The common tail off in network growth indicates that despite differing information seeking approaches the identification and integration of new sources of information declines significantly in the second half of the recorded sessions. This implies that the participants were beginning to switch to evaluating or applying the information rather than searching in the latter half of the session. Although this may be desirable in some cases a specific requirement in this study was that the participants spend the whole session seeking new and useful information. As such, if sustained seeking activity is too be encouraged it is suggested that stimuli or other interventions are considered to open up new information sources or break possible fixation. The main limitation of this study was the small sample size. Although this does affect the certainty with which the results can be applied the small sample was considered appropriate given the explorative nature of the study.

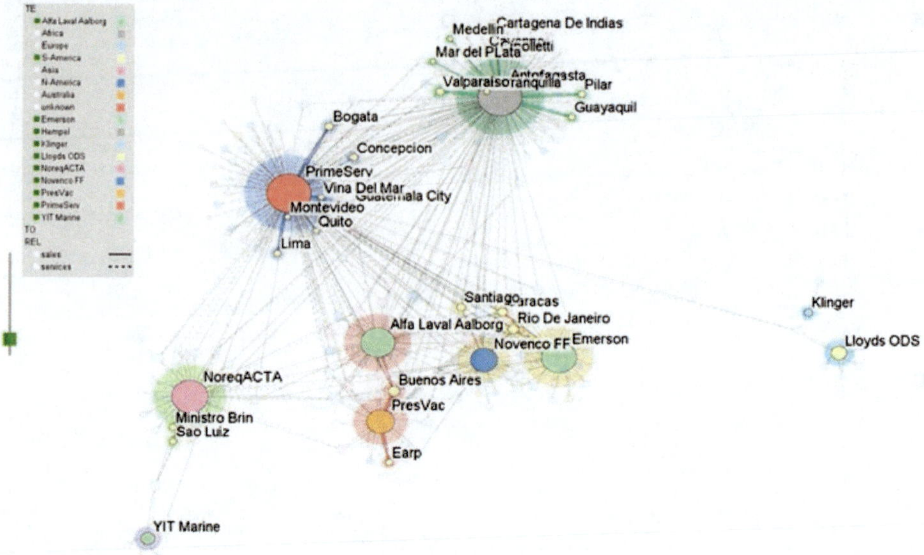

Fig. 3 Analysis of service providers according to geographic location

4.3 *After-Sales Services for Supporting Open Innovation*

The main goal of the PROTEUS Innovation Consortium (McAloone et al. 2011) is to develop new knowledge about how after-sales service can be effectively integrated into business development and industrial organisations, so as to become a source of revenue, rather than a cost to the company. The PROTEUS consortium consists of 10 companies of various sizes, which operate as suppliers in the Danish maritime industry branch. In order to understand how to effectively and systematically integrate service development into product development and business creation processes, the OrganicViz tool was employed to visually analyse qualitative data relating to the distribution network collected from the maritime partners (Fig. 3).

In this case the OrganicViz tool was used to show affinities, differences, gaps and opportunities for the consortium. To start with, two specific networks were created and analysed.

- The network of companies, related by the type of service provided. This was used to assess the evolution of the structural communities of the companies that share the same interest and to provide consortium members the opportunity to plan collaboration in the future.
- The network of locations, where services or sales are provided by each company. This was used to understand the evolution of the geographical location distribution for each specific type of service provided by members of the consortium and to enable them to find gaps and opportunities in specific geographic regions.

5 Discussion and Conclusions

Accordingly to the definition provided by Arthur (2009), information evolution is driven by dynamic information transformation process where all information objects composed of information fragments are related by ties of common descent from the collection of other information objects. In studying information evolution, there is a need to take into account the dependencies between informational content and context and the cognitive dynamics in order to systemically link information evolution to knowledge creation, learning resistance, information overflow, selective processing and innovation.

The way in which the information is processed is necessarily influenced by the organisational dynamics of processes which created and consumed that very information. Information dynamics as the process that encompasses the generation, dissemination, filtering, reprocessing and storage of particular types of information across different types of hierarchical social networks involved in complex R&D organisations must be viewed holistically with a systemic perspective, bearing in mind the complexity of the fundamental design process.

Therefore, the general science of networks and its various multi- and trans-disciplinary applications such as visual analysis have significant relevance for engineering systems design research. As it was illustrated by the case studies, it could be important for understanding the structure and behaviour of complex socio-technical systems, workload balancing in engineering design processes or analysing different types of interactions between stakeholders during the PSS life cycle. Network thinking of the kind described in this paper could be fundamental for developing new and effective techniques for solving the problems in the engineering design research related to the interpretation of the huge amount of data captured during experiments and observations that are more and more used as a main research method.

It is also worth highlighting that understanding information processing in complex natural or social networks could inspire novel approaches in our research field, as has been illustrated in our case examples. Information processing in such systems emerges from multiple feedback mechanisms, and allows the system to generate and use the right resources at the right place and right time in order to benefit the entire system. All this is done without central control, in a continually changing environment and in the face of multiple, often conflicting, requirements. If applied to contemporary engineering systems design processes it enables a paradigm where behaviour patterns which enhance a system's ability to adapt successfully could be identified, stabilized and applied to improve the development of organisations and product/service-systems in a changing environment existing on the global market.

The main limitation of the presented approach is in linking the patterns of network evolution to effective metrics. Performed studies have highlighted areas where such metrics might be developed but have not implemented them in the assessment process itself. As such, this would pave the way for further investigation and provide a basis for development of the visualization tool. Dynamic network analysis as an emergent scientific field applied to the engineering design research field, could potentially offer new understanding of the phenomenon that turn out during the engineering projects.

References

Ahmed S., Štorga M.: Merged Ontology for Engineering Design: Contrasting an empirical and a theoretical approach to develop engineering ontology, Cambridge University Press, USA: AIEDAM—Artificial Intelligence for Engineering Design, Analysis and Manufacturing, Vol. 23 No.4, p. 391–407 (2009)

Arthur, W.B.: The Nature of Technology, Free Press (2009)

Braha, D., Bar-Yam, Y.: The Statistical Mechanics of Complex Product Development: Empirical and Analytical Results. Management Science, Vol. 53 No. 7, pp. 1127–1145. INFORMS (2007)

Braha, D., Maimon, O.: A Mathematical Theory of Design: Foundations, Algorithms, and Applications. Kluwer Academic Publishers, Boston, MA (1998)

Cash, P., Stanković, T., Štorga, M.: "An Analysis of Engineers Information Seeking Activity, Proceedings of the ASME 2013 International Design Engineering Technical Conferences & Computers and Information in Engineering Conference IDETC/CIE 2013, Portland, Oregon, USA (2013)

Collins, S.T., Bredely, J.A., Yassine, A.A.: Analysing product development task networks to examine organisational change. IEEE Transactions on Engineering Management, Vol. 57, No. 3, pp. 513–525 (2010)

Dong, A., Moere, A.V.: Visualising Collaboration in Very Large Design Teams. Proceedings of the 15th International Conference on Engineering Design—ICED'05, Melbourne, The Design Society, Australia, (2005)

Fry, B.J.: Visualizing Data, O'Reilly Media, Inc. (2008)

Klein, M., Sayama,H., Faratin, P., Bar-Yam, Y-.: The dynamics of collaborative design: Insights from complex systems and negotiation research. D. Braha, A. Minai, Y. Bar-Yam, eds. Complex Engineered Systems: Science Meets Technology. Springer, New York, 158–174 (2006)

Li H., Xu, Z.: Analysing core-periphery structure of complex product development using social networks analysis. IEEE International Conference on Business and E-Government (ICEE 2011), (2011)

McAloone, T.C., Mougaard, K., Neugebauer, L.M., Nielsen, T.A., Bey, N.: Orthogonal views on product/service-system design in an entire industry branch. In Proceedings of 18th International Conference on Engineering Design, ICED11, Copenhagen, Denmark, Vol. 4, (ISSN: 2200–4342) pp. 77–87 (2011)

Mihm, J., Loch, C.H., Huchzermeier, A.: Problem-solving oscillations in complex projects. Management Sci. 49(6) 733–750 (2003)

Mitchell, M.: Complex Systems: Network thinking. Artificial Intelligence, Vol. 170, pp. 1194–1212. Elsevier (2006)

Mostashari, A.: Sociotechnical Systems: A Conceptual Introduction. In: Hu, F., Mostashari, A., Xie, J. (eds.): Socio-Technical Networks Science and Engineering Design. CRC Press, Taylor & Francis Group (2010)

Newman, M. E. J.: The structure and function of complex networks. SIAM Review 45, 167–256 (2003)

Ogawa, M., Ma K.L.: code_swarm: A Design Study in Organic Software Visualization. IEEE Transactions on Visualization and Computer Graphics, Vol. 15 Iss. 6, pp. 1097–1104 (2009)

Pavković, N., Štorga, M., Bojčetić, N., Marjanović, D.: Facilitating Design Communication through Engineering Information Traceability, AIEDAM: Artificial Intelligence for Engineering, Design, Analysis and Manufacturing Vol. 27, pp. 91–105 (2013)

Stanković, T., Štorga, M., Stojić, I., Savšek, T.: Traceability visualisation toolkit. In Proceedings of the 12th International DESIGN Conference DESIGN 2012, FSB, Zagreb, The Design Society Glasgow (2012)

Štorga M., Pavković N., Bojčetić N., and Stanković T.: Traceability of Engineering Information Development in PLM Framework. In Proceedings of the 8th International Conference on Product Lifecycle Management—PLM'11, Eindhoven, The Netherlands (2011a)

Štorga, M., Marjanović, D., Savšek, T.: Reference model for traceability records implementation in engineering design environment. In Proceedings of the 18th International Conference on Engineering Design—ICED 11, Copenhagen, Denmark (2011b)

Yassine, A., Braha, D.: Complex concurrent engineering and the design structure matrix method. Concurrent Engrg. 11(3) 165–176 (2003)

Selecting Models from Biology and Technical Product Development for Biomimetic Transfer

Helena Hashemi Farzaneh, Maria Katharina Kaiser, and Udo Lindemann

Abstract When searching for innovative solutions in technical product development, engineers increasingly search for analogies in other disciplines. Biology provides a large reservoir of solutions developed during the evolution with a high potential for technical analogies. Still, different approaches in biology and technical product development as well as a different terminology can hinder an effective transdisciplinary knowledge transfer. Models similar in biology and technical product development can serve to overcome this communication barrier. In this work, we collect and analyze models from both disciplines to select suitable models for a knowledge transfer from biologists to engineers and vice versa. The selection is based on an analysis of information needs in technical product development and biology. This is a first step for developing a "map" of models which provides paths for communication between both disciplines in transdisciplinary collaboration.

Keywords Transdisciplinary collaboration · Biomimetics · Knowledge transfer

1 Introduction

Is biomimetics—the application of knowledge of "living systems" in research and development approaches to solve technical problems and develop technical inventions and innovations (VDI 2012)—a transdisciplinary design approach?

Both mechanical engineering, including technical product development, and biology as the science of "living systems" are disciplines, i.e. have "a specific body of teachable knowledge with their own background of education, training, procedures, methods and content areas" (Apostel 1972). The term *transdisciplinarity* describes the transgression (Nicolescu 2002) or dissolution of boundaries between disciplines (Apostel 1972; Beneke 2003). According to Häberli et al. (2001), the goal of *transdisciplinary collaboration* is to solve complex problems of society. It involves a tendency to comprehensive exchange between the disciplines (Beneke 2003). Ertas

H. H. Farzaneh (✉) · M. K. Kaiser · U. Lindemann
Institute of Product Development, Technische Universität München, München, Germany
e-mail: hashemi@pe.mw.tum.de

L. Blessing et al. (eds.), *The Future of Transdisciplinary Design*,
https://doi.org/10.1007/978-3-319-06382-9_6

(2010) emphasizes that transdisciplinary collaboration creates new, shared knowledge. The Oxford dictionary explains the prefix *trans-* by *through, across, beyond* (Oxford University Press). Accordingly, the knowledge created in transdisciplinary collaboration should reach beyond the knowledge of the specific disciplines. This is the case for biomimetics: In a transdisciplinary design process, knowledge from biology and technical product development is merged to design new, innovative technical products, thereby creating transdisciplinary biomimetic knowledge.

To be able to create this biomimetic knowledge, engineers and biologists first have to gain an understanding of the other discipline's knowledge—a knowledge transfer is required. Still, barriers, such as different manners of presenting information and terminology hinder this knowledge transfer (Helten et al. 2011; Jordan 2008).

How can the transfer of knowledge between both disciplines be facilitated?

Probst et al. (2010) defines knowledge as cross-linked information and information as data which is set into context. Following this definition, knowledge comprises information and knowledge transfer requires information transfer. Information can be implicit (Nonaka and Takeuchi 1995) and undocumented, explicit (Nonaka and Takeuchi 1995) and documented in an unstructured form (e.g. text Kohn et al. 2010) or documented in a structured form in representations, i.e. models. The structured representation in these models facilitates the handling and consequently the transfer of information and knowledge.

In previous work, similar models have been identified in biology and technical product development (Schenkl et al. 2010). Still, the information needs of engineers developing technical products and biologists performing research are different. We hypothesize that the different information needs also reflect in the use of models in the two disciplines.

Therefore, in this work, the information needs of engineers developing technical products and of biologists conducting research are analyzed. Based on the information needs, product development and biology models containing relevant information are selected.

2 Literature Review: Using Models for Biomimetic Transfer

In Sect. 2.1, a review on literature to biomimetic transfer depicts two different approaches to the actual transfer of information between biology and product development. To clarify the understanding of models in both disciplines, an overview on model definitions is given in Sect. 2.2.

2.1 Biomimetic Transfer

As the transfer of information between biology and product development is a central challenge of biomimetics, a number of researchers have addressed the information transfer. Their procedures, methods and recommendations can be clustered into two main approaches.

The first approach particularly focusses on the transfer from biology to product development. To provide the product developer an access to biological solutions, (Hill 1997; Gramann 2004) and (The Biomimicry 3.8 Institute) developed taxonomies of functions and map them to biological systems. Löffler (2009) included further biological classification criteria for his taxonomy. The taxonomies provide access to information about biological systems that has been edited and inserted in catalogues Hill (1997), Gramann (2004) or databases (The Biomimicry 3.8 Institute) Löffler (2009). The information includes a short description of the biological systems (The Biomimicry 3.8 Institute) Löffler (2009), photos or sketches Hill (1997), Gramann (2004), (The Biomimicry 3.8 Institute), Löffler (2009), an abstracted biological principle Löffler (2009) or performance parameter Löffler (2009) for example.

Instead of using function-based taxonomies, Chakrabarti et al. (2005) and Vattam et al. (2010) developed biomimetic models which allow a mapping to biological systems on several concretization levels. Both the SAPPhIRE model Chakrabarti et al. (2005) and the DANE model Vattam et al. (2010) are based on technical constructs of function, behavior and structure. Information about a technical and a biological system has to be modeled by persons who are familiar with the constructs. Then, a technical construct and the corresponding biological construct can be mapped to transfer information.

To conclude, the first approach uses constructs from technical product development to provide access to biological systems. Therefore, information about the biological system has to be modeled from a technical perspective. The supported information transfer direction is from biology to technical product development and the expected users are engineers.

The second approach urges the inclusion of the biology perspective. This is achieved by a direct collaboration of engineers and biologists Helten et al. (2011) or discussions moderated by an intermediate person who is familiar with both disciplines Jordan (2008). Helten et al. (2011) emphasizes the importance of the clarification of the different terminology and "mind-sets" of each discipline. Jordan (2008) developed a semantic wiki to facilitate the search for discussion partners in both disciplines. In difference to the first approach, no structuring or modeling of information is performed to facilitate the information transfer between the two disciplines.

2.2 *Model Definitions*

A general definition of models is given by Stachowiak (1973) who defines a model as a representation of an original. It has a reduction feature, i.e. it does not represent all of the original's attributes and a pragmatic feature, i.e. it is used instead of the original for a specific purpose at a certain time Stachowiak (1973).

In technical product development, the concept of *product models* is used in accordance with this definition by Ponn and Lindemann (2011), Vajna et al. (2009) and Kohn et al. (2012) as a representation of a product (or another product model). It serves a certain purpose (pragmatic feature) and is simplified in comparison with the product (reduction feature) (Ponn and Lindemann 2011; Vajna et al. 2009).

In biology, Haefner (2005) defines models as descriptions of systems. *Description* implies the reduction feature. As pragmatic feature he names three possible purposes: Understanding, prediction and control. According to Leonelli (2008) a model is a partial representation of a phenomenon or the theory applied on the phenomenon (pragmatic feature). A model is a "rendering", i.e. a transformation or abstraction of the phenomena Leonelli (2008).

In conclusion, both the product development and the biology definitions of models are similar to the general model definition by Stachowiak (1973).

3 Approach

As described in the previous section, existing approaches for biomimetic transfer provide the engineer with a model of specific biological systems (first approach) or focus on a direct interaction of engineers and biologists without providing a model-based support for information transfer (second approach). The aim of this work is to identify models in both disciplines to provide both the engineer and the biologist with transferred models which serve their information needs in biomimetic collaboration.

According to Helten et al. (2011), in biomimetic collaboration, engineers and biologists directly interact in discussions. Moreover, engineers perform product development activities and biologists research activities. It is assumed that, to perform these activities, both have information needs which must be fulfilled. Therefore, the engineer's activities in technical product development and the biologist's activities in research are regarded. With regards to engineers, Ponn and Lindemann (2011) list technical product development activities. On the biologist's side, the activities of the research process model (RPM) according to Leedy (1989) are analyzed. The information needs of engineers are identified by analyzing their activities (Sect. 3.1), the biologists' information needs by analyzing the activities according to the RPM and additionally a biologic organization model (BOM) based on Campbell et al. 2008) (Sect. 3.2).

3.1 *Engineers' Information Needs in Technical Product Development*

With regards to engineers, Ponn and Lindemann (2011) have listed specific technical product development activities according to *requirements*, *functions*, *working principles* and *embodiment* of a technical product. The activities for which biomimetic transfer can play a role are selected and depicted in Table 1. Examples for activities for which biomimetic transfer is not relevant are acquiring customer's requirements, search for relevant norms and laws, searching catalogues of physical effects etc. The engineer's information needs for each activity are deduced from the description of the activity provided by Ponn and Lindemann (2011) (see Table 1).

As described in the introduction, structured information represented in models facilitate the handling and transfer of information. Therefore, biological models containing the information have to be assigned to the engineer's information needs for each activity. To facilitate the assignment of biological models, the engineer's information needs are summarized in 9 categories as shown in Table 1.

To select the biological models containing the required information, a wide spectrum of sources can be regarded, from general biological textbooks, books on specific biological issues to reviews on biological research to research publications. Regarding general biological textbooks (for biology students) on the one end of the spectrum, they provide an overview on all fields of biology and include a wide range of biological models. On the other end of the spectrum, biological research publications focus on a very specific topic, but include the latest research results. As the aim of this work is to transfer up-to-date information from biological research, two hypotheses are raised:

- Hypothesis 1: Biological models containing the required information are included in general biological textbooks. Biological researchers are familiar with the models and can model their research object using them.
- Hypothesis 2: Biological models containing the required information are included in research publications. They can directly be used for transfer to the technical product development domain.

In Sect. 4, we test both hypotheses, select biological models containing the information and assign them to the engineer's information needs.

3.2 *Biologists' Information Needs During Research*

The RPM based on Leedy (1989) is shown in Fig. 1. It describes six stages of the research process: problem identification, problem clarification, division of the problem into subproblems, building of hypotheses, collection of facts and confirmation/rejection of the facts. The process is cyclic, i.e. after the last stage, new problems are identified (first stage). In which of these stages and for which specific

Table 1 Product development activities according to Ponn and Lindemann (2011) and corresponding information needs

	Activity	Information need	Category n°
Requirements	Define requirements of different requirement types	Requirement types	1
	Define requirements for specific life cycle phases	Life cycle phases	2
	Define requirements by comparing existing technical systems	Properties[a] of different biological systems	3
	Define requirements by analyzing system elements and their relations	System elements and their relations	4
	Cluster requirements according to requirement type	Requirement types	1
Functions	Build a functional model by decomposing the main function into partial functions	Main and partial functions	5
	Define the system boundaries	Functional model	5
	Analyze existing systems by examining the functions of their elements	Functions of system elements	5
Working principles	Analysis of the relevant relations by geometrical decomposition	Working principle areas, embodiment design	6
	Analysis of the relevant relations by sequential decomposition	System states, sequences	7
	Evaluation of the working principle properties	Working principle properties	3
	Structuring and combination of working principles according to the partial functions	Assignment of working principles to partial functions	8
	Specification and systematic variation of Working principle areas	Working principle areas	6
	Evaluation of the overall working concept properties	Working concept properties	3
Embodiment	Decomposition of the system into subsystems according to functions	Assignment of system elements to partial functions	8

(continued)

Table 1 (continued)

	Activity	Information need	Category n°
	Decomposition of the system into subsystems according to interfaces	Interfaces (mechanical, geometrical, material/energy/ signal flow)	4
	Design for force flow and cascading of forces	Embodiment design	6
	Variation of embodiment designs	Alternative embodiment designs	9
	Evaluation of the embodiment design	Embodiment design properties	3
	Design of surfaces	Surface design	6
	Design of the interfaces	Interfaces (mechanical, geometrical, material/energy / signal flow)	4
	Evaluation of the overall embodiment model	Embodiment model properties	3

[a]According to (Ponn and Lindemann 2011), an object's property comprises a characteristic and its value

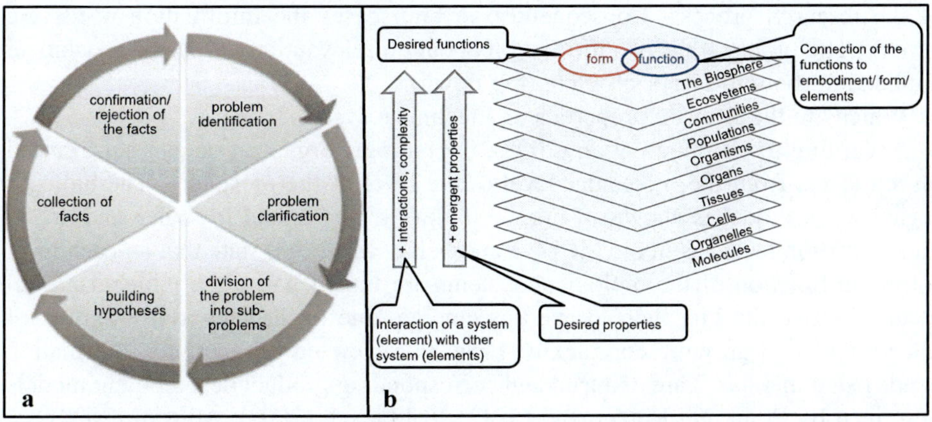

Fig. 1 Research process model (RPM) based on Leedy (1989) and Biologic organization model (BOM) based on Campbell et al. (2008)

activities do biologists need information from the engineer in a biomimetic collaboration? Considering two distinct starting points for biomimetic projects, the bottom-up approach (for a biological "solution" a technical application is searched) and the top-down approach (to solve a technical problem, a biological solution is searched), two different activities are identified.

1. Bottom-up: A biologist has identified a (part of a) biological system or phenomenon with potential for application in the technical domain, and searches

Table 2 Biology information needs and corresponding product development models

Biology information need	Product development model (Ponn and Lindemann 2011; Pahl et al. 2007; Adunka 2007)
Interaction of a system (element) with other system (elements)	Relational system model
Desired properties	Requirements list
Desired functions	List of functions, different functional models
Connection of the functions to embodiment/form/ elements	Sketches of working principles, morphological box TRIZ functional model

for technical applications. This activity is performed at the end of the last stage of the RPM, when the hypotheses about the biological system or phenomenon are confirmed or rejected.

2. Top-down: A biologists explores a biological system which has potential for solving a technical problem. This activity is performed at the beginning of the first stage of the RPM, when the biologist identifies a biological problem.

As the RPM is cyclic, both the last and the first stage refer to the same moment of the research process. Consequently, in both cases the information needs are similar: The biologist needs information about all relevant properties of a technical application (bottom-up) or problem (top-down).

Which are the relevant properties of a technical system?

According to Campbell et al. (2008), the biologic organization can be regarded on ten levels from the molecular level to the level of the biosphere. The biologic organization model is shown in Fig. 1. Its levels correspond to scales and not to concretization levels as in the MCM. Campbell et al. (2008) state that on each level form and function of the biological systems are linked. From the molecular level to the level of the biosphere properties emerge, due to the increasing number of interactions and growing complexity. From this view on biology, the information needs listed in Table 2 are deduced and corresponding product development models described by Ponn and Lindemann (2011), Pahl et al. (2007), Adunka (2007) are assigned.

4 Selection of Biological Models

To test the two hypotheses raised in the last section, general biological textbooks (hypothesis 1) and research publications (hypothesis 2) are analyzed. For hypothesis 2 semi-structured interviews with biological researchers are performed to achieve a more detailed understanding for the use of models in different research areas.

4.1 Hypothesis 1

Two general biological textbooks (Campbell et al. 2008; Purves et al. 2004) are chosen which are used in the first semesters of university biology courses worldwide as they have been translated into a number of languages. Therefore, it is assumed that most biologists are familiar with one of these textbooks and the contained models. The two textbooks were analyzed by the authors and similar models were identified in both textbooks in several sections treating different areas of biology. The similarity is explained by the models depicting development cycles of two biological organisms: gametophytes (the haploid phase of plants producing egg cells/sperms) in Purves et al. (2004), p. 751, and plasmodium (a single-cell parasite causing malaria) Campbell et al. (2008), p. 583. Despite the different objects of representation (gametophytes/plasmodium) and the different biology book, both development cycles are similar: They contain the same elements (annotated sketches, arrows) and have the same structure (the annotated sketches of different stages of development connected by arrows). Content wise, both represent relations between different system elements and the change of a system. Based on the content, six different model features are identified. All analyzed models possess at least one of these features:

- Relational feature: The model represents relations between several biological systems or system elements
- Change feature: The model represents the change of a biological system or system elements
- Morphological feature: The model represents the morphology of a biological system, i.e. the shape and/or relations of its elements
- Comparative feature: The models represents a comparison between several variations of biological systems or its elements
- Data feature: The model represents data acquired about a biological system or its elements.
- Mathematic feature: The model is a mathematic representation of a biological system or its elements.

Biological models which address a specific engineer's information need have a common model feature. Table 3 shows the assignment of one exemplary model from each biology book to one information need according to their common feature. It has to be noted that one model can possess several features and can therefore address several information needs. Examples are the models of gametophytes and plasmodium which possess the relational and the change feature and contain information regarding the life-cycle phases and the system elements and their relations.

Each model feature was identified in several models in both biological textbooks. As both biological textbooks contain similar models covering the engineer's information needs, hypothesis 1 is verified.

Table 3 Biological models assigned to the information needs of technical product development

Category n°	Information need	References: biological textbooks (one example per book)	References: publications (all models are referenced)	Common model feature
1	Requirement types	Campbell et al. (2008) p. 1231, Purves et al. (2004) p. 436		Relational feature
2	Life cycle phases	Campbell et al. (2008) p. 583, Purves et al. (2004) p. 751	Wedlich-Söldner et al. (2003) Fig. 1D, 3A	Change feature
3	Properties of different biological systems/working principle/working concept/embodiment design/embodiment model properties	Campbell et al. (2008) p. 1088, Purves et al. (2004) p. 946	Wedlich-Söldner et al. (2003) Figs. 1C–E, 2B, 3B–E, Geist and Auerswald (2007) Figs. 2–7, Eggers et al. (2011) Fig. 2, Yu et al. (2010) Figs. 1–3, Table 1, Michler et al. (2009) Figs. 3, 7–9, Table 3, Grimmler et al. (2011) Figs. 1–6, Foitzik et al. (2011) Figs. 1–3, Tables 1, 2, Corfield et al. (2011) Figs. 2–6, Fischer et al. (2009) Figs. 1–6	Data feature
		Campbell et al. (2008) p. 1230, Purves et al. (2004) p. 1042	Michler et al. (2009), Fischer et al. (2009) (various equations)	Mathematic feature
4	System elements and their relations, interfaces	Campbell et al. (2008) p. 581, Purves et al. (2004) p.342		Morphological feature
		Campbell et al. (2008) p. 189, Purves et al. (2004) p. 886		Relational feature

(continued)

Table 3 (continued)

Category n°	Information need	References: biological textbooks (one example per book)	References: publications (all models are referenced)	Common model feature
5	Main and partial functions, functional model	Campbell et al. (2008) p. 845, Purves et al. (2004) p. 975	Michler et al. (2009), Fig. 1, Grimmler et al. (2011) Fig. 7	Change feature
6	Working principle areas, embodiment designs, surface designs	Campbell et al. (2008) p. 115, Purves et al. (2004) p. 660	Wedlich-Söldner et al. (2003) 1a, b, f, g, 2a, c–j, Corfield et al. (2011), Baeumler et al. (2008) Fig. 1, Baeumler et al. 2008) Figs. 3–6	Morphological feature
7	System states, sequences	Campbell et al. (2008) p. 841, Purves et al. (2004) p. 910		Change feature
8	Assignment of system elements/working principles to functions	Campbell et al. (2008) p. 681, Purves et al. (2004) p. 853		Morphological feature
9	Alternative embodiment designs	Campbell et al. (2008) p. 529, Purves et al. (2004) p. 686	Beck et al. (2005) Fig. 1	Comparative feature

4.2 Hypothesis 2

To gain an overview on the use of models in biological research and their use to represent information in biological publications, semi-structured interviews with 11 biological researchers (professors and post docs) were conducted. To cover the biological discipline as complete as possible, biologists from different areas of biology are chosen for the interviews, such as ecology, physiology, evolutionary biology, theoretical biology etc. In the interview, one of the biology researcher's publications containing the models the biologists indicated as relevant was chosen for discussion. The discussion provided insights for the subsequent analysis of the models contained in the publications.

The result of the analysis is shown in the right column of Table 3. In this case, references for all models are given, not only for one exemplary model as was the case for the textbooks. It can be seen that five out of the nine engineer's information needs are not addressed by any of the 11 publications. Moreover, out of the 79 models identified in the publications, 62% (49) have the data feature, 24% (19) possess the morphological feature and merely 45% (5) of the publications contain models with other properties.

Both the data and the morphological feature represent merely data and not information (data set into context according to Probst et al. 2010). In addition, the maximum number of engineer's information needs addressed by models in one publication is three (in Wedlich-Söldner et al. 2003; Michler et al. 2009).

Despite the fact that this analysis was not performed on a sufficient number of publications to provide quantitative results, it can be concluded that a significant number of publications do not include more than one or two model features. In particular, a significant number of models used provide data, but not the necessary information about the biological system. This information is provided in unstructured form, i.e. in the text of the publications. Consequently, hypothesis 2 is disproved.

5 Discussion

There are a number of restrictions of this work. In particular, this work solely presents a first step towards understanding the biologist's information needs and his understanding of models: In contrast to the technical product development side whose activities have been analyzed, for example by Ponn and Lindemann (2011), the analysis of the RPM and BOM can only provide first hypotheses as to the biologist's information needs. These hypotheses can be tested by surveys with biologists performing biomimetics for example. As to the biologist's model understanding, the models and their features identified in biological textbooks provide options for modeling biological information, but it has to be tested if biologists can actually model their research results with these model features.

6 Conclusion and Outlook

In this work, technical product development and biological models were selected for supporting a biomimetic transfer. The selection is based on an analysis of information needs of the engineer's activities in technical product development (Sect. 3.1) and the activities of the biologist (Sect. 3.2). The engineering models containing the information requested by the biologists have been described by Ponn and Lindemann (2011), Pahl et al. (2007), Adunka (2007) and are listed in Table 2. For the identification of the relevant biological models containing the information requested by the engineer, two hypotheses are tested: The first hypothesis is verified: General biological textbooks contain the relevant biological models. Moreover, common features of the models addressing a specific information need are identified. Biologists who are familiar with the textbooks and model features can use them for modeling biological systems. The second hypothesis is falsified: In a number of publications, models are not (or to a small extent) used for representing information. Instead, the information is represented in unstructured form. Therefore, for a transfer of the latest research results, a modeling step has to be performed and information has to be modeled using the model features identified in biological textbooks (see Table 3).

In addition to addressing the limitations discussed in Sect. 4.2, in a next step, both technical and biological models will be analyzed in detail with regards to their content, structure and language. Moreover, different specifications of the models will be identified (e.g. flow oriented and relation-oriented functional modeling). The analysis will provide information for a "map" of models. Similarities will be identified in this map and serve to deduce communication paths between the disciplines. Using these communication paths, models can be transferred to the other discipline to support the transdisciplinary collaboration of biologists and engineers in biomimetic development projects.

References

Adunka, R., *Die Funktionsanalyse—eine Methode für die Konzeptfindungsphase (Lecture).* (2007): Erlangen

Apostel, L., *Interdisciplinarity. Problems of Teaching and Research in Universities.* Organisation for Economic Cooperation and Development: Paris (1972)

Baeumler, N., Haszprunar, G., and Ruthensteiner, B.: 3D interactive microanatomy of Omalogyra atomus (Philippi, 1841) (Gastropoda, Heterobranchia, Omalogyridae). Zoosymposia. vol. **1**, pp. 101–118 (2008)

Beck, A., Ritschel, A., Schubert, K., et al.: Phylogenetic relationships of the anamorphic genus Fusicladium s. lat. as inferred by ITS nrDNA data. Mycological Progress. vol. **4**(2), pp. 111–116 (2005)

Beneke, F.: Produktentwicklung. Arbeiten in und mit verschiedenen Disziplinen—wozu? In: F. Brand, F. Schaller, and H. Völker (eds.) Transdisziplinarität. Bestandsaufnahme und Perspektiven, pp. Universitätsverlag Göttingen, Göttingen (2003)

Campbell, N.A., Reece, J.B., Urry, L.A., et al.: Biology. Pearson, San Francisco (2008)

Chakrabarti, A., Sarkar, P., Leelavathamma, B., et al.: A functional representation for aiding biomimetic and artificial inspiration of new ideas. Artificial Intelligence for Engineering Design, Analysis and Manufacturing (AIEDAM). vol. **19**, pp. 113–132 (2005)

Corfield, J., Kubke, M.F., Parsons, S., et al.: Evidence for an Auditory Fovea in the New Zealand Kiwi (Apteryx mantelli). PLoS ONE. vol. **6**(8) (2011)

Eggers, S., Parks, M., Grupe, G., et al.: Paleoamerican Diet, Migration and Morphology in Brazil: Archaeological Complexity of the Earliest Americans. PLoS ONE. vol. **6**(9) (2011)

Ertas, A.: Understanding of Transdiscipline and Transdisciplinary Process. Transdisciplinary Journal of Engineering and Science. vol. **1**(1), pp. 55–73 (2010)

Fischer, B., Taborsky, B., and Dieckmann, U.: Unexpected Patterns of Plastic Energy Allocation in Stochastic Environments. Am Nat. vol. **173**(3) (2009)

Foitzik, S., Fröba, J., Rüger, M.H., et al.: Competition over workers: fertility signalling in wingless queens of Hypoponera opacior. Insectes Sociaux. vol. **58**, pp. 271–278 (2011)

Geist, J., and Auerswald, K.: Physicochemical stream bed characteristics and recruitment of the freshwater pearl mussel (Margaritifera margaritifera). Freshwater Biology. vol. **52**, pp. 2299–2316 (2007)

Gramann, J.: Problemmodelle und Bionik als Methode. Dr. Hut, München (2004)

Grimmler, C., Janssen, H., Krauße, D., et al.: Genome-Wide Gene Expression Analysis of the Switch between Acidogenesis and Solventogenesis in Continuous Cultures of Clostridium acetobutylicum. Journal of Molecular Microbiology and Biotechnology. vol. **20**, pp. 1–15 (2011)
Häberli, R., Bill, A., Grossenbauer-Mansuy, W., et al.: Synthesis. In: J. Thompson Klein, et al. (eds.) Transdisciplinarity: Joint Problem Solving among Science, Technology and Society—An Effective Way for Managing Complexity, pp. Birkhäuser Verlag, Basel—Boston—Berlin (2001)
Haefner, J.W.: Modeling biological systems. Springer, New York (2005)
Helten, K., Schenkl, S., and Lindemann, U.: Biologizing Product Development—Results from a Student Project. In: International Conference on Research into Design, pp. 27–34, Research Publishing,Bangalore, India (2011)
Hill, B.: Innovationsquelle Natur: Naturorientierte Innovationsstrategie für Entwickler, Konstrukteure und Designer. Shaker, (1997) ISBN: 3826528875
Jordan, A.: Methoden und Werkzeuge für den Wissenstransfer in der Bionik. Otto-von-Guericke-Universität Magdeburg, Magdeburg (2008)
Kohn, A., Peter, G., and Lindemann, U.: The Challenge of automatically annotating solution documents. In: KEOD 2010—International Conference on Knowledge Engineering and Ontology Development, INSTICC Press,Valencia, Spanien (2010)
Kohn, A., Lutter-Günther, M., Hagg, M., et al.: Handling Product Information- Towards an Improved Use of Product Models in Engineering Design. In: International Design Conference DESIGN 2012, pp. 1719–1730, Dubrovnik, Croatia (2012)
Leedy, P.D.: Practical Research. 4 ed., Macmillan Publishing Company, New York (1989)
Leonelli, S.: Performing abstraction: two ways of modelling Arabidopsis thaliana. Biology and Philosophy. vol. **23**(4), pp. 20 (2008)
Löffler, S.: Anwenden bionischer Konstruktionsprinzipe in der Produktentwicklung. Logos Verlag, Berlin (2009) ISBN: 3832521542
Michler, F., Eckhorn, R., and Wachtler, T.: Using Spatiotemporal Correlations to Learn Topographic Maps for Invariant Object Recognition. Journal of Neurophysiology. vol. **102**, pp. 953–964 (2009)
Nicolescu, B.: Manifesto of transdisciplinarity. State University of New York Press, Albany (2002)
Nonaka, I., and Takeuchi, H.: The Knowledge-Creating Company—How Japanese companies create the dynamics of innovation. Oxford University Press, New York (1995)
Oxford University Press, https://oxforddictionaries.com
Pahl, G., Beitz, W., Feldhusen, J., et al.: Konstruktionslehre. 7 ed., Springer Verlag, Heidelberg (2007)
Ponn, J., and Lindemann, U.: Konzeptentwicklung und Gestaltung technischer Produkte. 2 ed., Springer, Heidelberg (2011)
Probst, G., Raub, S., and Romhardt, K.: Wissen Managen—Wie Unternehmen Ihre Wertvollste Ressource Sinnvoll Nutzen. Gabler, Wiesbaden (2010) ISBN: 978-3-8349-1903-8
Purves, W.K., Sadava, D., Orians, G.H., et al.: Life: The Science of Biology. 7 ed., Sinauer Associates, Sunderland (2004)
Schenkl, S., Kissel, M., Hepperle, C., et al.: Communication Paths in Biomimetic Design—supporting a model-based interdisciplinary information transfer. In: NordDesign 2010, Göteborg (2010)
Stachowiak, H.: Allgemeine Modelltheorie. Springer, Wien (1973)
The Biomimicry 3.8 Institute, https://www.asknature.org/article/view/biomimicry_taxonomy
Vajna, S., Weber, C., Bley, H., et al.: CAx für Ingenieure. Berlin (2009)
Vattam, S., Wiltgen, B., Helms, M., et al.: DANE: Fostering Creativity in and through Biologically Inspired Design. In: T. Taura and Y. Nagai (eds.) Design Creativity 2010, Springer, London (2011)
VDI, *VDI 6220: Conception and Strategy—differences between biomimetic and conventional methods/products*. Verein Deutscher Ingenieure, Beuth Verlag: Berlin (2012).
Wedlich-Söldner, R., Altschuler, S., Wu, L., et al.: Spontaneous Cell Polarization Through Actomyosin-Based Delivery of the Cdc42 GTPase. Science. vol. **299**(21), pp. 1231–1234 (2003)
Yu, Z., Kastenmüller, G., He, Y., et al.: Differences between Human Plasma and Serum Metabolite Profiles. PLoS ONE. vol. **6**(7) (2010)

Transdisciplinary Design Approaches

Play and Transdisciplinary Understanding

Poul Kyvsgaard Hansen, Ade Mabogunje, Morten Lund, and Louise Møller Nielsen

Abstract The paper explores how an approach that has similarity to children's play can support exploration and communication of complex problems. The most common perceptions of children's play are discussed and it is illustrated how these perceptions can be transferred to an adult play setup. It is illustrated how this can be applied in a specific organizational context and the implications of this are discussed. The experiences from the reported case are compared with a large number of cases and the findings are used to define five essential explanations for the efficiency of the method. These five explanations are summarized in five comprehensive reflection points. The reflection points are formulated in a form that both support the further practical development of the methods and a form the invites for further theoretical research.

Keywords Serious play · Game · Complexity

1 Introduction

An important aspect of managing innovation is the ability to assess, review, and challenge a number of relevant parameters and viewpoints associated with the competitiveness of the product or service. Several empirical studies emphasize that successful innovation is more likely to happen when multiple viewpoints are applied and are specifically impacting the final solution (Francis and Bessant 2005; Sawhney et al.

P. K. Hansen (✉)
Faculty of Engineering and Science, Aalborg University, Aalborg, Denmark
e-mail: kyvs@mp.aau.dk

A. Mabogunje
Center for Design Research, Stanford University, Stanford, CA, USA

M. Lund
Business School, Aalborg University, Aalborg, Denmark

L. M. Nielsen
Department of Architecture, Deign and Media Technology, Louise Møller Nielsen, Aalborg University, Aalborg, Denmark

L. Blessing et al. (eds.), *The Future of Transdisciplinary Design*,
https://doi.org/10.1007/978-3-319-06382-9_7

2006; VHA Health Foundation 2006). The ability to apply multiple viewpoints can be referred to as one of the most important Innovation Management functionality parameters, and the result can be measured as an essential part of the innovation capability of the organization. In essence, this multiple viewpoint ability is a competence that requires methods to support communication and synthesis across traditional organizational borders.

Perceptions of the same problems are often different in various internal and external organizational functions. So communicating these different perceptions requires supporting methods. Synthesizing is even more challenging because it involves the emergence of new solutions that integrate insights and disciplines from various organizational functions. The process of integrating insights is often supported by some degree of adaption or adoption of methods between disciplinary boundaries. In innovation processes the research of user behavior is often supported by various ethnographical methods. Ethnography was originally developed in anthropology but has been adopted and adapted by other disciplines and serves as an efficient cross-disciplinary method of researching and communicating user behavior and needs. These exchanges between disciplines can be defined as transdisciplinarity.

Innovation requires explorative activities. When solutions are not known it is essential to be able to facilitate a process that asks the essential question: What can be? The answer to this question requires the ability to be imaginative and creative. However, children's play includes some of the same elements, and in a transdisciplinary tradition it is relevant to ask whether the insights from children's play can support innovative processes among adults.

The simple answer is that insights from children's play can support innovation processes. However, due to most adults' view of play as a rather childish activity it is necessary to apply play in a thoroughly facilitated way. If facilitated efficiently, play can support communication in cross-disciplinary contexts, and, additionally, play can support the synthesis of different personal and organizational perceptions of complex problems.

2 Transdisciplinarity and Complexity

Most approaches to problem solving are informed by academic disciplines that emphasize insight into the classics of the discipline and a deductive reasoning approach (Robinson 2001). Historically, this type of approach has proven powerful from the time of the industrial revolution until present time. Dave Snowden argues that this perception of problems relates to what might be characterized as complicated problems (Snowden and Boone 2007). Complicated problems are problems in which the relationship between cause and effect can be revealed by analyses and application of expert knowledge. Decisions on how to deal with the problems will be based on the outcome of the analyses. More and more, the relationship between cause and effect is not easily determined, and in many cases, the logical explanation of the relationship only reveals itself while conducting retrospective analysis.

Complex problems are very different in nature compared to complicated problems. They are messier and more ambiguous; they are more interconnected with other problems; more likely to react in unpredictable non-linear ways; and more likely to produce unintended consequences (Buchanan 1992).

The approaches of how to deal with complicated and complex problems are also very different. When dealing with complicated problems, the approach is described as a sequence of generic types of activities: *sense-analyse-respond*. Sensing is referring to the use of all senses to identify a given problem. When identified it is realized that the problem needs analysis in order to understand the causal relationship between the problem and the various features that impact the problem. The analysis will then support the decision on how to respond. This is in line with the whole establishment of the present academic tradition (Robinson 2001). Most people are academically trained in various analysis techniques and these do to a large extent constitute the various academic disciplines.

When dealing with complex problems, the approach is described as a very different sequence of generic types of activities: *probe-sense-respond*. Since there are no immediate cause-effect relationships, there is a need to *probe* and *sense* whether this probe supports an explanation or provides a partial solution to the problem. If it supports it, it will be part of the subsequent *respond*.

The differences between the two approaches are significant. Given the fact that by far most people are trained in the tradition of complicated problems, there are numerous methods that support solving these types of problems. Some of the consequences are that complex problems are treated as complicated problems and this leads to solutions that will not solve the real problems.

In order to deal more effectively with the complex problems, new transdisciplinary approaches are needed. Merely adjusting the existing approaches cannot solve the problems. One of the fundamental needs is methods that support the communication between existing disciplines and thereby facilitates the emergence of transdisciplinary approaches. Recent research documents that play involves elements that support such emergences (Beck and Wade 2004).

3 Play in a Transdisciplinary Context

In most settings, play is associated with games, and in a developmental psychological perspective, these two phenomena fit well together. When children play, they are normally in a game setting—a game setting understood as a situation governed by certain rules and certain timing constraints. The game provides structure to the activity and the psychological state play.

The most common adult understanding of game has the same characteristics. A board game or a card game has certain rules and predefined ways of interaction. When we play cards, our interaction is strictly regulated by rules, and the psychological state of play is associated with the challenges that are built into the structure of the game. This view has led to a perception of play and game as primarily leisure activities that serve various social interaction purposes.

However, this perception involves a risk of reducing play to play by the rules. This is too narrow an understanding of play. And the broader understanding of play is also supported by the function of play in children's play.

Jean Piaget discovered that children are not just passive absorbers of experience and information, but active theory builders. Children are not just empty vessels into which we can pour knowledge. Rather, they are theory builders who can construct and rearrange knowledge based on their experiences in the world. His theory of knowledge, stipulating that knowledge is built or constructed by the child, is known as *constructivism* (Piaget 1951).

Seymond Papert extended the theory of constructivism to the fields of learning. Papert eventually named his theory *constructionism* (Papert 1996). It included everything associated with Piaget's constructivism, but went beyond it to assert that constructivist learning is especially present when people are engaged in constructing something external to themselves.

Papert extended the theory to cover the adult learning process as well as children's learning process. Constructionism is a way of making formal, abstract ideas and relationships more concrete, more visual, more tangible, more manipulative, and therefore more readily understandable. At the core of both ideas is the notion that when we "think with objects" or "think through our fingers", we unleash creative energies, modes of thoughts, and ways of seeing what most adults have forgotten they even possess (Papert 1996).

In line with Piaget and Papert, it makes sense to focus on play in the meaning of being playful rather than playing a game. Playfulness is seen as a psychological state that encourages curiosity and willingness to see and understand settings in different ways. This interpretation is highly relevant when focusing on the relevance of play among professional adults. Many complex problems require means that support organizations and individuals in the process of identifying new yet unknown possibilities or in the process of reframing the current views of a situation.

There is a further encouragement in this pursuit by additional research that reveals various relevant aspects of applying play when dealing with complex problems:

- The cognitive benefit of drawing on the imagination to develop new insight (Papert 1996).
- The social benefit of developing new frames for interaction (Vygotsky 1978).
- The support for sensemaking in the context of giving meaning to observations or experience (Sutton-Smith 1997).
- The emotional benefits of providing positive affective associations as well as a safe context in which to take risks, to try on new roles, and to explore new potential forms of practice (Bateson 1972).
- The cognitive benefit of deep concentration by loosening the sense of time resulting in an increased involvement (Csíkszentmihályi 1990).
- The benefits of using improvisation as a critical concept in terms of supporting team creativity (Gardner 2007; Kudrowitz and Wallace 2010).

The support by research to see play as a serious tool for business purposes is significant. However, the perception of play and games as mainly leisure activities is still dominant among most professionals. In order to overcome this resistance towards play and games the notions of "Serious Play" and "Serious Games" have emerged (Alvarez et al. 2010).

4 Serious Play

The notion of Serious Play and Serious Game is normally viewed as a recent phenomenon. According to Sawyer the game "America's Army" released in 2002 was the first successful and well-executed serious game that gained public awareness. Sawyer defines "Serious Games" as "any meaningful use of computerized game/game industry resources whose chief mission is not entertainment" (Sawyer 2007). Michael Zyda, who participated in the development of America's Army, proposes a similar definition: "A mental contest played with computer in accordance with specific rules that uses entertainment to further government or corporate training, education, health, public policy, and strategic communication objectives" (Zyda 2005). Both definitions represent the dominant view of Serious Games as being synonymous with computerized games. However, as stated above this is a too narrow definition that involves a risk of underestimating the important factor of playing or being playful in order to communicate complex insight or overcoming challenges.

Computerized games have proven powerful in order to support an understanding of the interplay between many parameters, but are generally less powerful in dealing with open-ended problems. Therefore, a broader or supplemental understanding of Serious Games and Serious Play is needed.

In his book "Serious Play" Michael Schrage applies a broad definition: "Serious play is about improvising with the unanticipated in ways that create new value. Serious play includes any tools, technologies, techniques, or toys that let people improve how to play seriously with uncertainties" (Schrage 2000). Schrage emphasizes that the essence of play remains the same while the methods, and, in particular, the bandwidth changes. As the bandwidth for play settings increases, the opportunities for richer interaction between people and their ideas do as well. But the fundamental phenomena involved in the play setting remain the same. Similarly, new tools for modeling, prototyping, and simulation may not change the meaning of innovation, but they can potentially change how organizations innovate.

Therefore, it is essential to consider in parallel the development of game technology and the organizations that utilize the technology. The most intriguing perspectives do not come from looking at what these new technologies can do, but from using them to see how people and their organizations behave and change their behavior.

5 Methodology

In order to focus more on the behavioral perspectives of serious games a relatively simple open-ended play setup—LEGO Serious Play (LSP)—has been chosen. This decision is based on the general applicability of the setup. In the past, LSP was only available to trained and certificated consultancy professionals. From June 2010, however, it has been made "open source".

In practice, LSP is a facilitated workshop where participants are asked different questions in relation to an ongoing project, task or strategy. The participants answer these questions by building symbolic and metaphorical models of their insights in LEGO bricks and present these to each other. An essential part of the LSP workshop is the non-judgmental, free-thinking and playful interaction between the participants (Gauntlett 2007).

The LSP process has four central elements:

Construct – Give meaning – Tell the story – Reflect

In a specific workshop, the participants are initially asked to build their perception of the pre-defined problem. The dogma of the process is 'Start building'. As the spontaneous building process progresses, the participants give meaning to the models by tapping into their brains. After the individual building assignment, each person is given time to explain his or her perception of the problem at hand with outset in the physical model. Other participants will ask about details, but will respect the model and the meaning that the individual builder attaches to it. This last part is the reflection part, which provides insight both for the individual and the team.

When each of the participants has told their stories the whole team builds a shared model based on elements from and additions to the individual models. The shared model can then be challenged in various ways according to the overall purpose of the workshop.

The elements of LSP are rather generic and have been used for a broad range of purposes, including: strategy development, communication, and exploration; strategic relations to market segments or partners; organizational development; innovation and product development; change management processes; scenario development and testing; mergers and acquisitions processes; branding workshops; leadership and team development; turnaround and restructuring; market entry; competitive analysis; value chain analysis; many specific applications within social sciences.

The applications exist in many variations and both the variations and the nuances to the applications are driven and developed by private consultants.

This research is conducted as a series of case studies that have been chosen on the basis of "realism". Realism in the sense that it has to be a real problem to the organization and that the play setting potentially is supposed to impact critical decisions regarding how to deal with the problem. These requirements have been defined in order to ensure that the participating organizations have a high level of motivation.

The whole portfolio of case studies comprises 70 + studies. This paper does build on the prior case studies but in order to exemplify one particular case is described in more details.

6 Case Study on Exploration of the Guiding Principles Through Play

The case company (ALFA) is a large Danish industrial company with global activities. They have been growing constantly in size and profitability for the past 60 years.

The continuous globalization of the company frequently challenges existing structures and procedures and the company is frequently developing and testing new methods to support the associated changes. Department A has the responsibility of documenting processes and methods, and they have, due to globalization, been growing and are currently feeling a strong need to review their setup and internal strategy. They see a possibility in applying a game/play approach to support this process in order to introduce new methods that can challenge existing practice.

The game session is discussed and planned during two meetings. At the first meeting Department A Management gives its perception of the challenges and the facilitators give input to how the game can be applied and what can be the potential output. Various theoretical concepts are discussed. The management has considered some concepts and some have been tested in other companies and are reported by the facilitators. After the first meeting, the facilitators develop a plan for the game and this is discussed in details at a second meeting.

In the specific case, it is agreed to plan for a six-hour game. The scope of the game is agreed to focus on the guiding principles that have been applied in Department A. Management feels that these have been defocused in the period with intensive growth and, furthermore, that many new employees are less confident with the current guiding principles. The facilitators initially suggest the overall theoretical concept. This is inspired by the ideas behind Toyota's way of operating and focuses on the relationships between guiding principles and the outcome (Liker 2004). Based on the guiding principles, methods are developed and the applications of these methods play a supporting role in obtaining the results. Among the managers of department A there is a strong feeling that the growth of ALFA has led to the development of a variety of methods that do not explicitly take into account the guiding principles. Furthermore, there is some doubt about what the guiding principles of the department are.

The specific game is organized in four rounds where the 28 participants are divided in smaller groups of 7:

1. The participants are taken through an introduction to the game method.
2. The participants each build three examples of perceived successes.
3. The participants build methods associated with these successful interactions.

4. Each group builds shared models of the perceived methods applied and discusses candidates for guiding principles.

During the first round, the participants are getting familiar with the LSP method. In five short exercises they learn the ideas behind LSP and are moving to a stage where they feel confident in expressing their ideas with LEGO models. First, they get familiar with building and then stories are attached to the models and shared with the participants in the small groups.

During the second round, the participants build their perception of successful interaction with their internal customers in other departments of the larger ALFA organization. The models and the rich stories behind are shared within the small groups. In total, the participants build 80+ examples that are documented with pictures and notes.

In the third round each participant builds models of the methods that have been applied in each of their own perceived success stories. As in the previous session, these personal and detailed models of the methods applied are shared within the smaller groups. The personal perceptions of the methods applied provide detailed experiences and these experiences become platforms for lively discussions within the groups. This illustrates that the individual participants in some cases have rather different views of the method parameters.

In the third round, the participants are asked to build their individual models of their perceptions of the applied methods into shared models. These models have a richness highly impacted by the differences of the individual models. The groups now propose candidates for the guiding principles that inform the methods, and, finally, all groups share their results.

Pictures and video clips of the stories document the most important parts of the whole session.

7 Reflections

The reflections are based on 70+ different workshops that in general only share the usage of the LEGO bricks and the few overall guidelines provided by the LSP method.

The documentation varies in rigor. Some workshops have been fully video documented and some workshops have been documented only with selected pictures, smaller video clips, and notes from the workshop (see picture example in Fig. 1).

The purpose and the initial agreements with the sponsors have been documented and in 1/3 of the cases there have been conducted interviews with the sponsors and some of the participants after the workshops.

The different scopes of the workshops have made it difficult the to design formal evaluation methods so until now the interviews have been made with a very open agenda focusing on their perception of the process, their evaluation on the use

Fig. 1 Picture from a LSP workshop

of LEGO bricks, and their perception of the applicability of the outcome of the workshop.

The following is a first-hand aggregation of the experiences seen from both the facilitators' viewpoint and from the participant's viewpoint. The emphasis has been concentrated on aggregating the empirical observation to five insights that can drive and focus the future research activities.

7.1 Empirical Observations and Aggregation of Insights

When participants enter the room and see the vast amount of LEGO bricks, they immediately get play associations. They get strong associations to their childhood and their own children. At the same time they feel some reluctance: Can playing with LEGO be serious and can it support a serious purpose?

During the session, the reluctance disappears during the first round where the participants step by step gain confidence in expressing complex ideas with their LEGO models and the associated stories. When the participants after the first session are asked whether they feel confident about expressing any idea, question, concept, dilemma etc. with LEGO models, they always answer positively. The 70+ workshops provide no examples of people reacting negatively.

As the session progresses, the participants gain more and more confidence. The critical element is whether the problem at hand has been formulated precisely enough. During the whole session, the primary role of the facilitator is to monitor continuously the progress and to critically review whether the questions formulated and initially agreed on with the sponsors drives the discussion in the right direction. If the progress and the direction are right, the role of the facilitator often changes towards being mostly timekeeper. The teams are able to monitor the direction by themselves and to reframe the questions if needed.

The reflections on the experiences with the LSP setup can be summarized in five points:

1. The facilitation of collective exploration
2. The collective concentration and focusing
3. The emergence of a shared and sufficient language
4. The emergence of collective understanding
5. The emergence of individual and collective commitment.

7.2 *The Facilitation of Collective Exploration*

The initial process of each participant building a model representing his or her personal perception of the defined challenge, and the subsequent step of telling the story behind the model, has two purposes. First, the participants experience that they can express their perception with surprisingly many details. Second, the team members get to hear each participant's perception. In particular the more introvert-oriented participants benefit from being able to express themselves. New perspectives and uncertainties on details drive needs and wants for further exploration. Initially, the facilitator proposes further explorations of details or perspectives. But as the team becomes more confident, they themselves take over. The facilitator then acts mainly as timekeeper.

The game perspective is defined by the simple rules that apply: The participants build and take turns in telling their stories. This is amplified by the playfulness that applies due to the engagement in the building process and the vast amount of LEGO bricks with strong colors and many functions.

7.3 *The Collective Concentration and Focusing*

Concentration in general and in particular collective concentration in a meeting or a workshop setting is a major challenge. The experienced concentration in a LSP session can partly be explained by the playfulness that the participants experience. Another major influence is the complex interplay between the fingers and the brain. The fingertips are extremely sensitive and send a large amount of sensory information to the brain. This means that when people engage with their fingers, they generally tend to be more concentrated because of the large amount of sensory input sent to the brain (Wilson 1998).

The collective focusing is achieved by the framing of the initial problem or challenge. Most often, this is negotiated with the sponsor before the game. But in some cases the team itself does the best possible framing based on its pre-understanding. In these cases, the team simply builds a LEGO model of its perception of the relevant problem and the exact formulation is then negotiated within the team.

7.4 The Emergence of a Shared and Sufficient Language

Engaging in transdisciplinary problems requires some kind of shared language. In many organizations, this shared language is business English. Though most professionals feel confident in expressing themselves in English, it often lacks nuances for people that do not have English as their first language. When combining the spoken language with the 3D models in LEGO, participants experience that this supports them in expressing their personal perception of complex issues. Drawings would potentially have the same effect, but many people express less confidence in their own drawing capabilities. Therefore the LEGO models are not only a supplement to the spoken language but also a democratic means that allows everybody to express themselves on equal terms.

7.5 The Emergence of Collective Understanding

The LEGO models built by the participants can be seen as 3D metaphors that represent complex problems. Each comprehensive LEGO model is a metaphor in itself, but the many functional LEGO bricks also have metaphoric potential that is used when the participants tell their rich stories associated with the model. Metaphors are powerful in communication and are of cognitive importance as well (Lakoff and Johnson 1980). The game setup forces metaphoric thinking. First, on the person who builds and gives meaning to the model. Second, when telling the story, and third, on the people on the team when listening to the story.

7.6 The Emergence of Individual and Collective Commitment

In the initial process of getting confident with the LSP method, the participants are asked to pick 10 random LEGO bricks. Then they are told to build a tower as high as possible with the chosen bricks in 2 min. Hereafter the facilitator tells the participants that their towers have to be tested by swinging a wooden hammer towards them. Naturally, all towers crash.

The participants are asked how they felt when their towers were crashed, and they always respond that they had some emotional reaction to the incident. The facilitator then summarizes the experience by stating that they had developed emotional feelings about an insignificant model that they had 2 min to build.

A plausible explanation for the emotional reaction from the participants is that they have taken a sort of ownership to the model. The same ownership feelings apply when the models later on in the session represent something meaningful and important. This applies both to the individual models and to the collective models.

8 Conclusions

The paper illustrates how a play approach can support cross-organizational communication and facilitate transdisciplinary development. When combining a playful building process with rich storytelling people can improve their own insight into complex problems and become able to communicate this to relevant persons with a different background. It is illustrated how the setup of the specific LEGO Serious Play game by nature is related to the organization's relevant context and current challenges. To date the setup has been tested in 70+ different organizational settings, and though the settings have been different in nature the same simple process applies: building and telling rich stories. Though there is a level of reluctance when people are presented with the method, they consistently gain confidence within half an hour and hereafter they excel with the method. So far it has proven that facilitation is necessary. The differences in communication compared to normal written and oral communication are too significant for people to choose the method in spontaneous communicative situations.

The difference in empirical setup has until now set limits to the possibilities of applying more scientific rigor in the research process. Therefore these conclusions can be considered only as qualitative insights. However, the five elements in the LSP process that have been found in the reflection part of this paper will open for a more detailed and scientifically thorough research.

Acknowledgements The theme of the paper is associated with the EU Network of Excellence project GaLA—Game and Learning Alliance—EC Grant Agreement Number 258169. The research is partially funded by this project.

References

Alvarez, J., Alvarez, V., Djaouti, D., Michaud, L.: Serious Games: Training and Teaching–Healthcare—Defence and Security—Information and Communication. IDATE, France (2010)

Bateson, G.: Steps to an Ecology of Mind, Northvale NJ: Jason Aronson Inc. (1972)

Beck, J.C. and Wade, M.: Got Game—How the Gamer Generation is Reshaping Business Forever, Harvard Business School Press (2004)
Buchanan, R.: Wicked Problems in Design Thinking. Design Issues, Vol. 8, No. 2, Spring, pp. 5–21 (1992)
Csíkszentmihályi, M.: Flow: The Psychology of Optimal Experience, New York: Harper & Row (1990)
Francis, D., Bessant, J.: Targeting innovation and implications for capability development. Technovision 25, pp. 171–183 (2005)
Gardner, H.: Five minds for the future, Harvard Business School Publishing (2007)
Gauntlett, D.: Creative Explorations—New approaches to identity and audiences, Routledge, New York (2007)
VHA Health Foundation: The Power of Innovation, VHA Inc. (2006)
Kudrowitz, B., Wallace, D.: The Play Pyramid: A Play Classification and Ideation Tool for Toy Design. The Journal of Arts and Technology. 3(1), pp. 36–56 (2010)
Lakoff, G. and Johnson, M.: Metaphors We Live By, University of Chicago Press (1980)
Liker, J.: The Toyota Way: 14 Management Principals from the World's Greatest Manufacturer, McGraw-Hill Education (2004)
Papert, S.: The Connected family, Atlanta: Longstreet Press (1996)
Piaget, J.: The Child's Conception of the World, London: Routledge (1951)
Robinson, K.: Out of Our Minds—Learning to be Creative, Capstone Publishing Limited (2001)
Sawhney, M., Wolcot, R.C., Arroniz, I.: The 12 Different Ways for Companies to Innovate, MIT Sloan Management Review, Spring, 47(3), pp. 75–81 (2006)
Sawyer, B.: The "Serious Play" Landscape. Presented at the Instructional and Research Technology Symposium for Arts, Humanities and Social Sciences, Camden (2007)
Schrage, M.: Serious Play—How the World's Best Companies Simulate to Innovate, Harvard Business School Press (2000)
Snowden, D.J. and Boone, M.: A Leader's Framework for Decision Making, Harvard Business Review, November, pp. 69 (2007)
Sutton-Smith, B.: The ambiguity of play, Cambridge, MA: Harvard University Press (1997)
Vygotsky, L.S.: Mind in Society: The development of higher psychological processes, Cambridge MA: Harvard University Press (1978)
Wilson, F.:The Hand: How Its Use Shapes the Brain, Language, and Human Culture, Pantheon Books (1998)
Zyda, M.: From visual simulation to virtual reality to games. Computer 38(9), 25–32 (2005)

The Elusive Character of Design Ideas

Marc Tassoul and Petra Badke-Schaub

Abstract During design processes, interim results, i.e. ideas, are imagined, elaborated and shared in an on-going reflective creative conversation. One can argue that designers have an educational role in building a shared understanding with clients and other stakeholders, across disciplines of what the design is and why it is what it is. Especially in the context of trans-disciplinary collaboration, it is essential to reflect on what it means to collect and capture 'the whole story', i.e. not just the designed object, but also the context and the choices made in terms of what is of importance or essential for a resulting design.

Building such a shared understanding consists of the three following steps:

1. Inventorizing and building an understanding of the system
2. A prioritization of what is of essence, what is of importance,
3. Defining a direction for seeking solutions.

We call this 'information set' (understanding, essence and direction) the Design Vision. Related to this view, we explore media used to depict or articulate such a vision. It will be argued that the choice of representation (text, metaphor, sketch, etc.) is highly influential on the richness and usefulness of the collected information and the following creative process.

Keywords Design practice · Experiential knowledge · Human centered design

1 Introduction

In the context of design education, it can be observed how students and alumni gradually learn to envision an ever 'widening' system to be taken into account in their design projects, integrating users, user-product interaction, meaning, (intended) behavior, and socio-cultural aspects in their designs. Lawson and Dorst (2009) describe this as a developmental path from novice designer to the most accomplished visionary

M. Tassoul (✉) · P. Badke-Schaub
Faculty of Industrial Design Engineering, TU Delft, Delft, The Netherlands
e-mail: m.tassoul@tudelft.nl

L. Blessing et al. (eds.), *The Future of Transdisciplinary Design*,
https://doi.org/10.1007/978-3-319-06382-9_8

designer. The 'designed object' itself is but the tip of an iceberg of meaningful connections, on the one hand with users (and other stakeholders), a user context, and on the other, some organization(s) which supports (initiates, realizes, carries) this product or service.

The design process (i.e. the ongoing structure of planning and procedures, including methods, tools and techniques) and the design thinking of the designer (i.e. reasoning, attitude and skills) are claimed to be central in any design process, independent of its application, to name but a few: Product Innovation, Industrial Design Engineering, Social Innovation and Service Design.

During the design process there may be moments of illumination, e.g. in generating a core idea on which to base the enfollowing development process. However, taking into account a wide set of objectives and conditions, considering different options, taking decisions at varying levels of abstraction, and building a meaningful 'image', e.g. of a product in relation to an envisaged context, is not a simple thing. The nature of the unclear goal setting with the need for problem definition brought about the notion of design problems as "ill-defined" problems. (Reitman 1965; Simon 1973). Nowadays, complexity has even increased by additional requirements coming forth both from the context in which our products have to function and from the trans-disciplinary and multi-cultural collaboration in which these designs come to be.

During design processes, interim results are continually being generated and pictured by the design team in an on-going reflective creative conversation. At certain moments, results are shared between experts in various disciplines and with clients and potential future users to involve them in decision-making and idea generation. One can argue that designers have an educational responsibility in building a shared understanding on what a design is developing into, and why the design is what it is. And, depending on the background of these stakeholders (i.e. knowledge, interest, functional role), care about how the communication in pictures, words and other media—may have to be adapted to the specific role of these stakeholders.

But the 'design story' starts before a design problem is formulated, in a phase sometimes called the "Fuzzy Front End of Innovation" (Koen et al. 2001, Tassoul 2011). The outcome of this phase would be a story containing an overview and understanding of a context (1), a set of objectives (values and qualities) (2) and sometimes core ideas for solutions or at least directions for seeking solutions (3). In this paper, we focus on the first instance of a final design making its first appearance, usually in some rudimentary or implicit form, but demonstrating a basic structure and some of the vocabulary and grammar that will be further elaborated and detailed during an enfollowing design process. It is the moment when earlier analysis and problem definition have come to fruition in some synthesis, maybe into what one may now also call the Raison d'être of the design or 'Design Vision'.

Related to such a design vision and the acknowledgement of complexity and richness in understanding a contextual system, a frame and a tool are presented to help depict and articulate such a vision. The idea being that the media used to transfer such rich knowledge (in the form of text, tables, sketches, metaphors, etc.) will to a lesser or greater extend reduce the detailed character of the information itself, and thereby influence to what extent information can be articulated and shared between colleagues and other stakeholders in a multidisciplinary context.

2 Exploration

Analysis—A design process is generally started by collecting information on the contextual system and the building of an understanding of such a system including social and human-product interaction, leading to discerning key values and qualities which are then summarized in a Design Brief.

As van der Lugt and Visser (2005), referring to Sanders and William (2001), propose, we have to "make a distinction between research for validation and research for inspiration." For designers it is essential to get as close as possible to the actual context or user environment. Any abstracted (or 2nd hand) information can be detrimental to a proper understanding. Any reporting of information will per definition contain an element of reduction of the original observations, in part depending on the medium used to do such recording and communication.

Requisite Variety—Using Ashby's law of requisite variety as an analogy (Ashby 1956) we argue that the media used in representing context and ideas need to provide a sufficiently wide 'dimensional space' to allow for an optimal richness and diversity of information from 'the world', i.e. towards providing perspective (context), a proper problem understanding and definition (design brief), and be an inspirational source for idea generation and concept development.

Although addressing leadership, Lord et al. (2011) state "Requisite complexity is a complex adaptive system's concept that pertains to the ability of a system to adjust to the requirements of a changing environment by achieving equivalent levels of complexity."

This principle of requisite variety also seems to be applicable on the media used when capturing, archiving and communicating something like a Design Vision. When sharing information in the context of a collaborative design process, one should pay special attention at optimizing this transfer, thereby respecting as much as possible the richness and complexity of the original observations and ideas.

Design expertise—Lawson and Dorst (2009), referring to Dreyfus (2003), present a hierarchy distinguishing levels of development of designers ranging from (1) Novice—(2) Advanced beginner—(3) Competent—(4) Expert—(5) Master to (6) Visionary designer. An important distinction in defining such levels is the scope of the system envisaged by the designer, how much diversity and complexity are taken into account in the unfolding design.

Nigel Cross in his studies on 'Outstanding Designers' (Cross 2002) describes three key design activities that appear to be common to the expert designers studied in this paper: (1) taking a broad systems approach to the problem, rather than accepting narrow problem criteria; (2) 'framing' the problem in a distinctive and sometimes rather personal way; and (3) designing from 'first principles'.

Many efforts in Design Education are directed to motivate or at least introduce students to envisage such a wider system view, gradually allowing for more complexity and richness in their design processes and confront them with elements specific to 'higher' levels of apprenticeship as described earlier on (Lawson and Dorst 2009).

Design Process—At the outcome of a design process, there will be some tangible object one can use and test on all the aspects and criteria defined earlier on in the process and documented in a Design Brief. As a frame of reference, one could argue that in the stages preceding this stage, there will only be partial representations of a 'product-to-be'. And the further back one goes, the more vague, implicit and undefined ideas and their representations are. But this outcome doesn't stand by itself, it includes a Raison d'être, an argumentation why it is what it is (or should become). This argumentation finds its roots in a context (simply put, on the one hand users in a community and a society and on the other a business or other organization in a market) and in turn in an understanding of what is of importance, i.e. what are the main qualities one is after in developing an idea into a design.

Design Vision—Following our own research results (e.g. Tassoul 2011) we conclude that this vision consists of the following three fundamentals:

1. An overview and understanding of the system in which the future product is to function;
2. A prioritization of what is of essence, what is of importance, identifying the most important functions, qualities and values, including a prioritization between these different objectives;
3. A direction for seeking solutions; depending on timing (along a design process time line) these solutions can be more or less developed and elaborated from basic idea to actual product-in-use.

We argue that these three fundamentals form the major part of the 'whole story' around a design, in other words a 'Raison d'être' of a design.

Essence—"*De gustibus non est disputandum*" or "In matters of taste, there can be no disputes". Quite contrary to this idea, Steve Jobs, in his "Lost Interview" (Jobs 1995) states "Ultimately it comes down to Taste". This notion of taste may very well be the essence of what the design should be about, but at the same time be one of the most elusive elements. We have come not to argue about it, or negotiate around it to avoid such discussions.

This aspect of taste may very well be one of the main bottlenecks or at least a potential pitfall of trans-disciplinary collaboration, or any collaboration for that matter. At Apple, there would be a clear leading figure taking harsh decisions when needed. In more democratic, bottom-up or emerging processes, such avoidance of discussion, and lack of good contextual understanding of what is of essence may at times very well lead to average or even mediocre outcomes.

In their paper "Product Integrity", Clarck and Fujimoto (1990) describe the design process leading to a new Honda Civic, the CRX, in which at some point the design got to be characterized by the metaphor "Pocket Rocket". This was after an extended analysis and argumentum on a future concept, after which all the designers and engineers chose this symbolic metaphor to designate what the design should be about. This process greatly helped to get everyone 'on the same page'. Especially considering the trans-disciplinary character of such an effort. The use of such abstract, symbolic metaphor, might be a way to capture some of that elusive quality of taste.

On Elusiveness itself—Next to dealing with "ill defined problems" and the necessity to work in the context of a more and more trans-disciplinary social context, we can now pinpoint a number of elements of elusiveness around design ideas:

1. The designed object is only the tip of an iceberg; the whole story is much more comprehensive, containing context, essence and (direction for) solution;
2. This comprehensive story can loose much of its richness while being transposed into media for sharing with other stakeholders;
3. The design of the object itself is part of an unfolding creative process. Any interim results will be a potential source of confusion in communication, especially when not associated with the 'whole story' or Design Vision;
4. Some of the essential qualities may be almost impossible to pinpoint objectively, let alone be measured.

3 Relating and Naming

As Van der Lugt and Visser (2005) state, a shift is taking place in the design community. Nowadays there is an increasing emphasis on gaining a rich understanding of the experience and context of the user and some envisaged future product use. One of the techniques they describe (while referring to Gaver et al. 1999) is Cultural Probes, a tool to collect rich insights from future users. Similarly, in the recently introduced VIP method (Vision In Product design—Hekkert and Van Dijk 2011), a core element in their design process is the analysis of a future context and the definition of an interaction as a starting point for the actual product design. With more recent Base-of-the-Pyramid applications and the use of Design Thinking in Social Innovation (Kandachar et al. 2011, Tromp 2013), the trend is that design is to be seen as creating an interaction between an object and people in a social context, and its integration into much more complex and encompassing (human, social, cultural, environmental) systems. To get a better grip on such needs in present design practice, one needs to explore this act of exploration and analysis more deeply.

3.1 Richness, Complexity and Reduction

Starting with a metaphor: When wandering through a forest, there is an infinite amount of possible observations, e.g. of plants, animals, but also simply shades of green, the blue sky visible through the leaves, reflections and light mirroring in water droplets, but also the soft underground under one's feet, the air temperature along one's skin, scents, etc. etc. Delving a level deeper, one might discover how some species appear at particular moments during the day or night. And over time, one might start to discover dependencies and interactions between various species, etc.

We argue that a 'real' understanding of such a system can only come forth from actually stepping into that system. It is about 'Relating' with a system, and whether one is able to 'catch' the more underlying, hidden or intricate dependencies and potentials might also depend on knowledge acquired in other instances, and through reading and conversations with specialists. Having one's own experience of a system (context) gives one a totally different and much richer realization of such a system (Fig. 1).

But this is only one side of a dichotomy (Fig. 2). Any analysis and representation of such a 'natural' system will imply a reduction into a subset of variables and lead to a more abstract view of the observed system. This side of the dichotomy, one might

Fig. 1 Forest— *Source* https://www.imageafter.com/

Fig. 2 Relating and naming

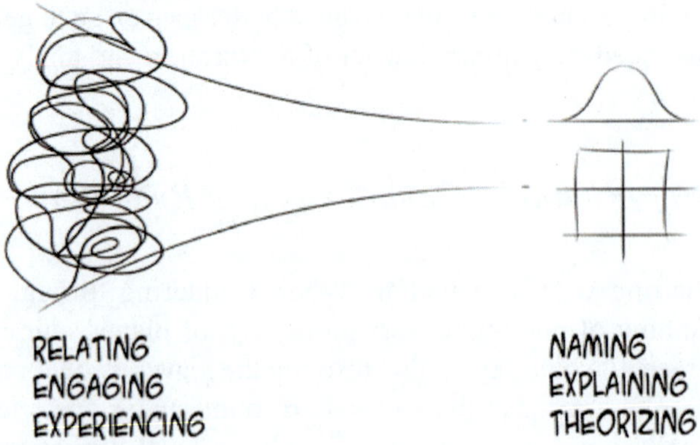

call Naming. This naming side is aimed at objective, measurable and generalizable descriptions; whereas the relating side is subjective, singular and anecdotal. Badke-Schaub and Dörner (2002), in their publication "Am Anfang war das Wort-oder doch das Bild-oder doch das Wort…" ("At the beginning, there was the Word, or was it the Image, or still the Word") in which they developed a reflection tool combining sketching and prototyping on the one hand and verbalising ideas on the other. This is kind of the same dichotomy. Sketches and prototypes help to imagine and develop ideas in the sense of "what if one were to…". Externalising an idea in the form of sketches and prototypes facilitates the 'Relating' idea. Verbalising ideas (Naming) facilitates a process of formalising ideas, and developing an argumentation in a more objective and measurable manner. What is important in a design context is to include both. William Gaver, in providing a description of Cultural Probes, notes (Gaver et al. 2004) that "the Probes embodied an approach to design that recognizes and embraces the notion that knowledge has limits. It's an approach that values uncertainty, play, exploration and subjective interpretation as ways of dealing with such limits." This model "Relating versus Naming" helps to put such an idea into perspective and provokes one to explicitly pay attention to both sides of this dichotomy. Also on Cultural Probes, Gaver states "Over time, the stories that emerge from the Probes are rich and multilayered, integrating routines with aspirations, appearances with deeper truths", and "The Probes simultaneously make the strange familiar and the familiar strange, creating a kind of intimate distance that can be a fruitful standpoint for new design ideas."

3.2 Saturday Special Edition

A tool used in various design projects at our faculty is the Saturday Special Edition, a newspaper format evoking a more journalistic, and maybe even anthropologist like attitude. As mentioned earlier, the medium has a great influence on the information contained. An analogy for this finding is Ashby's Law on Requisite Variety. Ross Ashby, scholar and author of "An Introduction to Cybernetics" (Ashby 1956) developed the idea that when regulating a system, a regulator needs to have sufficient 'degrees of freedom' to be able to react to the diversity of external events e.g. that might threaten a system. Similarly in design, if enough variety and richness is to be collected and communicated in some document, a traditional report might be extremely limiting. A more journalistic approach seems to be closer to what may be needed in exploring a context for a future design, allowing for a greater richness and diversity of information.

The content can consist of interviews with users, stakeholders and experts, cover relevant technologies, usage and customs, show the newest developments, State-of-the-Art applications, futuristic projections, Base of the Pyramid usage; it can also include adds, cartoons and comics, combining text, image, cartoon and many other media in one document.

The objective of using such a format is a response to finding traditional reports too limiting in terms of information collection and sharing. In traditional reports, information can easily be too abstracted to be of real value to a following design process. This format forces one to collect more first hand, rich information to Relate, Engage and Familiarize oneself with a design context. This might also be a fitting response to Sanders and William (2001) idea: “Make a distinction between research for validation and research for inspiration” (Fig. 3).

4 An Explorative Study

Preliminary to this study, a number of published interviews of designers were inventoried, like an interview with Bruno Ninaber van Eyben on the design of the Dutch Guilder and the Dutch Euro coins (Kwanten and Van de Ruit 2011). What some of these stories had in common was that they could each be brought back to the three main categories of information described earlier, namely (1) an overview and understanding of the system or context, (2) what is of essence (from specific qualities that the design should fulfill, to problem statements and criteria) and (3) a direction for solution finding.

To get a better grip on this idea of three categories of information, seven designers were invited to participate in this study. Four were interviewed individually and three in a focus group. The details can be found in Table 1. The table also includes information on whether mentioning user and context when asked about describing the design and design process.

The semi-structured interview consisted of questions which started by asking the interviewees to recollect a recent project, and then focus on the ‘first synthesis’ meeting, the interaction between themselves and their clients and other stakeholders and the information shared in this instance. Three kinds of questions were asked, starting with (1) open questions, then (2) focused open questions, and finally (3) closed questions. Examples of answers can be found in Table 2.

5 Analysis and Interpretation

One of the objects of study was the information which interviewees consider to be essential for sharing with clients and other stakeholders. Most respondents (4 out of 7) concentrated their answers on the designed object itself, on aspects like form-giving, technical elaboration, feasibility, cost, etc. Only after more precise questions, issues about user-product interaction and context came to the surface. When asked about this observation itself, the interviewees explained that the design projects mentioned had a ‘given’ design brief and therefore did not leave much space for one’s own exploration and (re-)definition of a design brief.

Fig. 3 Saturday special edition

Table 1 Respondents and mention of User/Context in design story

	Education	Years Experience	Mention 'Object'	Mention user/Context open questions	Mention user/Context half open and closed questions	Observations
1	Mechanical Eng. (M.Sc, Delft)	0	Yes	Yes	Yes	Interesting exception
2	Design Graduate (M.Sc, Delft)	1.5	Yes	No	–	Naming various reasons why not mentioning users and context: 1. Not involved in design brief definition 2. Different stage of design process
3	Design Graduate (B.Sc, Malaysia)	3	Yes	No	No	
4	Design Graduate (B.Sc EAFIT)	5	Yes	No	–	
5	Design Graduate (B.Sc EAFIT)	5	Yes	No	–	
6	Design Graduate (Pittsburg)	12	Yes	Yes	Yes	User/Context integral part of the "design story"
7	Design Graduate (M.Sc, Delft)	15	Partially	Yes	Yes	

However, the two most experienced designers (12 and 15 years) (see Table 1) do mention the user and the context right from the beginning. For example one interviewee designing a remote control for the elderly, made reference to what would be of importance to this user group. The other interviewee explicitly saw it his business to translate complex issues into representations of understanding and vision (in the sense of what is of importance) that can be shared with other stakeholders. Interestingly the Mechanical Engineering Master student, with relatively little experience, also made reference to users and user needs.

When developing a rationale behind these findings, a number of reasons may be identified: The amount of 'design freedom' one gets (or takes), and the stage at which a designer gets involved in a design process, has an influence on the scope of system view taken into consideration. Is it as early as the Fuzzy Front End of Innovation, during the early stages of a process when a design brief still has to be formulated, or is it much later during the elaboration and optimization of the design ?

Table 2 Examples of answers representing different categories of analysis

	Item	Answers
1	Observing and building an understanding of a system	"And what Mike does, he explores the space, finds all the elements and he structures them", " so, it's more like an overview of everything that they were doing", "Get your user to pump out all kind of information, find those patterns together"
2	Essence, values, qualities, objectives, criteria, design brief	"A good example how you get to a good story, and then being able to connect that to the essence of the company", "Then we move away from the session and actually try and find how the connection between the essence of something, and the structure of a larger thing, how those things are connected"; "in that sense I am always trying to find the essence of some things, or try to find the goal or the core of the problem, or the core of the story, and start to build from there", "Maybe this is the essence, and this is the structure"
3	Solutions and directions for solutions	Many examples of design solutions mentioned, but no metaphors (except for one respondent)
4	Managing the process	"It's usually a messy process to do with the client without having some expectation that are mismatching" "It's also very important to set some deadlines and define the resources that you need"
5	Media used	"This is a very visual process, but it's a conversation", "These conversations are very, they are so broad so abstract, conceptual even" "To make a visual communication that is able to convey what is happening here", "Or that can maybe have people connect with the message"
6	Interaction designer—Client	"And the client, yes, they were very happy, they praised us on the way that we went through the project with them, really linking at them as persons, as each separate person in a team and actually creating a story from them", "This whole fuzzy phase, for the company, this is hell,..., this is my comfort zone, it is all very recognizable for all people who are designers"

Another interesting reflection is that these findings are not contradictory (except for one respondent, the Mechanical Engineering Master student) to the model as proposed by Lawson and Dorst (2009) on levels of expertise. An important distinction being the scope of the system envisioned by the designer, how much diversity and complexity is to be taken into account in the unfolding design in relation to the amount of experience (here, measured in years of professional experience). A recommendation for further studies would be to run these interviews with more high-experience designers (>12 years of experience). But the selection of cases used as a tangible reference will have to be more precisely defined, e.g. cases in which the interviewee was also responsible for, or at least had a say, in the initial analysis and formulation of the design brief.

5.1 Building a Shared Understanding

Most respondents believe that the development of shared understanding was successful in the cases they mentioned during the interviews. However, some answers point to the risk of miscommunication in a "First Synthesis" meeting, especially with a larger number of stakeholders in the room. One interviewee referred to the strong discussions between different professionals when defining a hierarchy of values. Furthermore it was mentioned that there is a risk of information overflow when trying to build shared understanding. Another respondent stated that he aimed only at building enthusiasm for his idea/design, assuming that the client would not be able to understand what the design was about.

Some respondents mentioned (including the 3rd closed question phase) that this first synthesis meeting had been preceded by earlier meetings. In these meetings earlier design ideas and decisions had already been considered and most elements of the earlier mentioned design vision or Raison d'être had been shared and agreed upon. As such, assuming that a so-called 'First Synthesis' meeting can be pinpointed as a singular event along a time axis, might be somewhat artificial.

5.2 Design Vision

One respondent stated: "If I have a customer saying they want a rugged product, they know they want that. For me it can be like a tank, with rubber bumpers or external facets and different aesthetics and so I create a sort of vision of what that product may be with different applications of ruggedness and then if they say, I really like this sort of approach, then this becomes some sort of the vision." Here the vision seems to be more about the character and the expression of the object itself, what one might call a value, and how such a value should be materialized. No external reference (e.g. user context or other) was mentioned other than the client proposing such a product character. One can speculate on the reasons for such a limited representation of a

Raison d'être. It would seem that in a trans-disciplinary context, this might be a risky approach to adopt.

Another respondent stated on Design Vision: "It should be something like an abstract statement that you have to develop for your projects. Something in which you express what the context of use is, who would use it and how they will use it. A vision also has to be something that you state in time, a projection", a statement which reflects similar understanding as what was discussed earlier in this paper.

6 Conclusions and Recommendations

This research is about having a conversation between new ideas on design method, developed and applied during design projects in our faculty in recent years, and today's design practice through the voice of experienced designers.

The notion of "The Elusive Character of Design Ideas" refers to the process through which a design emerges from a rich and complex context. Consciously working with an idea of Design Vision or Raison d'Être during a design process may help deal with increasingly complex conditions and objectives today's products have to respond to. Also, within the context of trans-disciplinary projects, it seems useful to review which information ought to be collected and shared to be able to build a shared 'Frame of Reference'. This would be complementary to a more formal Design Brief and an operational set of Criteria, thereby acknowledging and respecting the richness and complexity of the original system for which one designs.

Some assumptions were not tested, e.g. by prescribing such a deliverable as the Saturday Special Edition, (student-) designers might be induced to take into consideration a wider and more detailed system view as described by Cross (2002), and thereby more quickly climb competence levels as described by Lawson and Dorst (2009).

With an increasing attention for sustainability, co-design and social innovation, we need tools to face this increasing complexity amidst which new products must be designed, products that respond to the demands of today's circumstances. The ideas presented in this paper on the "Relating versus Naming" dichotomy and the "Saturday Special Edition" as an alternative mean to record and communicate context information, and what a Design Vision could consist of in support of design process in a trans-disciplinary setting seem promising in achieving such goals.

Acknowledgements A word of thanks to Maria Hock and Joanne Cheung for assisting with the interviews and the data analysis, Joris Wondergem for the illustration, Helga Hohn for proof-reading, and Will McWhinney for one of the core ideas in this publication, "Relating versus Naming."

References

Ashby, W. R., An Introduction to Cybernetics. Chapman & Hall, London (1956)
Badke-Schaub, P., & Dörner, D., Am Anfang war das Wort—oder doch das Bild—oder doch das Wort. In W. Hacker (Hrsg.), Denken in der Produktentwicklung. Psychologische Unterstützung der frühen Phasen. Zürich: Rainer Hampp Verlag. S.27–52 (2002)
Clarck, K. B., & Fujimoto, T., The Power of Product Integrity, Harvard Business Review, Nov–Dec (1990)
Cross, N., Creative Cognition in Design: Processes of Exceptional Designers. In T. Hewett and T. Kavanagh (eds.) Creativity and Cognition, ACM Press, New York, USA (2002)
Dreyfus, H. L., The Spinoza Lectures. University of Amsterdam (2003)
Gaver, W., Dunne, T., & Pacenti, I., Cultural probes. ACM Interactions, 6(1) (1999)
Gaver, W. W., Boucher, A., Pennington, S., & Walker, B., Cultural Probes and the Value of Uncertainty. Interactions—Funology, 11(5), pp. 53–56; Goldsmiths Research Online (2004)
Hekkert, P., & van Dijk, M., Vision in Design, A Guidebook for Innovators, BIS publishers, Amsterdam, NL (2011)
Jobs, Steve, The lost interview, Magnolia Home Entertainment (1995)
Kandachar, P., Diehl, J. C., Parmar, V. S., Shivarama, C. K., Designing with Emerging Markets—Design of Products and Service, TU Delft, The Netherlands (2011)
Koen, P., Ajamian, G., Burkart, R., Clamen, A., Davidson, J., D'Amore, R., Elkins, C., Herald, K., Incorvia, M., Johnson, A., Karol, R., Seibert, R., Slavejkov, A., Wagner, K., Providing Clarity and a Common Language to the "Fuzzy Front End", Research-Technology Management, Volume 44, Number 2, 1 March 2001, pp. 46–55(10) (2001)
Kwanten, M., & Van de Ruit, J., A Decade of Design, The 10th anniversary of the Euro; Turn the Page #38, TU-IDE Delft, NL (2011)
Lawson, B., & Dorst, K., Design Expertise, Elsevier, Oxford (2009)
Lord, R. G., Hannah, S. T., & Jennings, P. L., A Framework for Understanding Leadership and Individual Requisite Complexity; in Organizational Psychology Review, Sage Publications (2011)
van der Lugt, R., & Visser, F. S., Widening involvement in Creative Group Processes, Proceedings of the 9th European Conference on Creativity and Innovation, Lodz, Poland (2005)
Reitman, W.R., Cognition and thought. New York: Wiley (1965)
Sanders, E.B.-N., William, C.T., Harnessing People's Creativity: Ideation and Expression through Visual Communication. In: (Eds. Langford, J., McDonagh-Philp, D.) Focus Groups: Supportive Effective Product Development. London: Taylor and Francis (2001)
Simon, H. A., The structure of ill-structured problems. Artificial Intelligence 4, 181–204 (1973)
Tromp, N., Social Design, How products and services can help us act in ways that benefit society, Doctoral Thesis, TU Delft, NL (2013)
Tassoul, M., Fuzzy Front End and Designing towards a Raison d'Être, in Roozenburg, N., L.L. Chen and P.J. Stappers (Eds.) Proceedings of IASDR2011, the 4th World Conference on Design Research; (pp. 1–12). Delft: TU Delft/IASDR. (TUD) (2011)

The Evolution of Systems Thinking in Interaction Across Disciplinary Boundaries—Cases from Space Industry

Hubert Anton Moser

Abstract Systems engineering is performed in multi-disciplinary teams involving disciplines such as finance, mechanics, and radiofrequency. This requires interaction across disciplinary boundaries. In such interactions systems thinking is learned. This evolution is not sufficiently understood. To improve this understanding, four cases in space industry were studied: a concept exploration project in a summer school, two concept exploration projects in a concurrent design facility, and five projects in several lifecycle stages in a small space systems company. A framework based on activity theory was developed to analyse work activity and its evolution with different temporal and organisational scales. Systems thinking is learned from seconds to years. Knowledge changes in two directions: vertically within disciplines, and horizontally across disciplines. A key factor for this change of knowledge is the multi-disciplinary quality of interaction. The major constituent of this quality, the awareness of the diversity and orientation towards extra-disciplinary interactors, is presented in detail.

Keywords Systems engineering · Systems thinking · Learning

1 Introduction

The work presented in this article is part of a research project described in Moser (2013c). The definitions introduced in this section resulted from an extensive literature study, details of which can be found in Moser (2013c). Systems engineering is an approach to product development applied during the product creation phase of the lifecycle, i.e. from concept exploration, via detailed development, production, until the use of the product in operation starts (Wertz and Larson 1999). A product (e.g. a space mission) is regarded as being related to a physical entity (Tan 2010), i.e. including physical products and services such as maintenance, consultancy, and operations.

H. A. Moser (✉)
Department of Microsatellites, LuxSpace Sàrl, Betzdorf, Luxembourg
e-mail: hubert.moser@gmail.com

L. Blessing et al. (eds.), *The Future of Transdisciplinary Design*,
https://doi.org/10.1007/978-3-319-06382-9_9

Systems engineering is: the management and engineering of a system, which is more than the sum of elements, applied throughout the lifecycle, involving perspectives from multiple disciplines, in a continuous iterative process (Moser 2013c).

Systems thinking is regarded as a prerequisite for systems engineering (Davidz and Nightingale 2008). Simplifying the definition of thinking as "active, persistent, and careful consideration of any belief or supposed form of knowledge in the light of the grounds that support it, and the further conclusions to which it tends" (Dewey 1997, p. 6), thinking is defined as doing something with knowledge (Shadrick and Lussier 2009). Systems thinking is defined as doing something with knowledge of components, context, relationships, and dynamics of a system-of-interest Moser (2013c). Systems engineering and systems thinking require humans in a work activity, i.e. humans in interaction.

1.1 Multi-Disciplinary Interaction Modelled as Boundary Management

Scholars from various research areas such as artificial intelligence, computer supported collaborative work, engineering design, sociolinguistics, and operations research have studied humans in interaction with humans and with technology. Bucciarelli (1984) and Schön (1983) are two of the first who considered engineering as a social activity and not as rational problem solving.

Boundary work is a model of human activity, which emphasises interaction between different contexts. Goodwin (1995) describes such an interaction with the term polycontextuality. Interaction within this polycontextuality is defined as multi-disciplinary interaction. The notion multi-disciplinary is preferred as it assumes neither collaboration between disciplines (inter), nor transcending of disciplinary boundaries (trans).

Differences between disciplines can be distinguished with the concept of epistemological distance. This concept describes "the extent to which an individual interacts with those with similar ways of knowing [...] or substantially different" (Adams et al. 2009, p. 344). Therefore, boundaries are described by significant epistemological distances and can be identified according to the epistemological distance that is displayed in interaction, i.e. through the behaviour of interactors.

Boundaries between disciplines and their related perspectives on work (e.g. mechanics, electronics, and radiofrequency) are one type of boundary Adams et al. (2009). Special types of disciplinary boundaries are those between lifecycle stages and their related disciplines (lifecycle disciplines such as concept designer, manufacturer, operator). In addition, the product under development often defines boundaries, e.g. by the system and subsystem boundaries or functional boundaries (Andreasen and Hein 2000).

Notions such as boundary zone (Gorman 2002), negotiation zone (Radford et al. 2006), and contact zone (Kramsch 1993) suggest regarding boundaries as something that is negotiated and maintained between participants. Different perspectives from different cultures meet, collide, and need to be managed. Therefore, the term boundary management is used to describe interaction between multiple contexts.

1.2 Learning as Evolution of Knowledge and Thinking

Across research areas, concepts of learning imply change as a major underlying concept. Learning is often defined as change of behaviour Pavlov and Gantt (1928). No statement about better or worse performance is included. Other definitions ("without assistance" (Hougaard 2009, p. 3), "more effectively" (Long et al. 2011, p. 14), "better" (Persidis and Duffy 1991, p. 254) indicate a widely shared perception where learning is regarded as change, which increases performance, hence making things better. Here, this is not regarded as necessary; learning is regarded as neutral change. Learning is an evolutionary process that can be described as change of behaviour, knowledge, perception, understanding, thinking, conception, perspective, and strategy.

Two directions on how systems thinking is expected to evolve have been identified (Lamb 2009). One direction is that systems thinking "results from the interaction of team members" and the second direction is that systems thinking is "an emergent behaviour" (Lamb 2009, p. 129).

To conclude, systems engineering is a multi-disciplinary activity that requires systems thinking. Systems thinking is expected to change within interaction of shorter and longer periods, i.e. systems thinking is expected to evolve within the systems engineering activity.

2 Approach

Following the suggested directions, this article concentrates on the research question: How does systems thinking evolve in multi-disciplinary interaction?

Answering the research question requires the study of the current situation in practice. Such an empirical study is required in order to identify factors which can be impacted to improve the current situation. The focus of this article is on this empirical study. In terms of the Design Research Methodology (DRM) (Blessing and Chakrabarti 2009), the focus is on the Descriptive Study 1. The interventionist parts of the research project (Prescriptive Study, and Descriptive Study 2), which focus on the improvement of the current situation, are described in Moser (2013c). Figure 1 gives an overview of the observation context of four case studies.

Study 1 (S1) was a ten-days-study in a European summer school. Four student teams comprising astrophysicists as users and engineers as technical officers worked

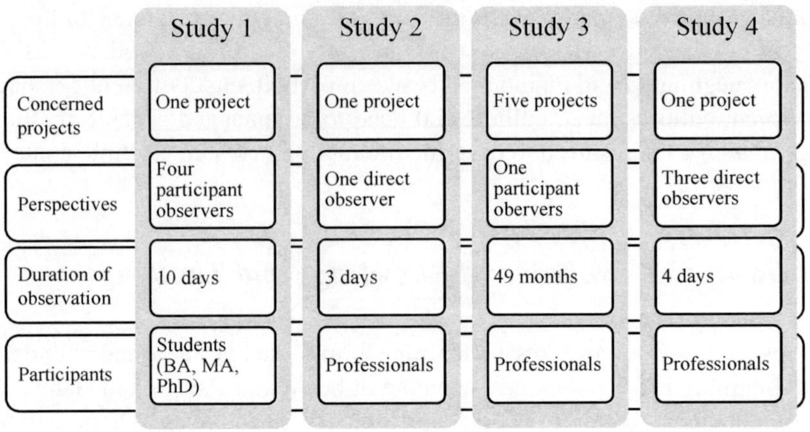

Fig. 1 Observation details of the four studies

in the concept exploration stage of a space mission. The author was a member of the teams.

Study 2 (S2) was a three-days-study in a concurrent design facility with an Earth observation space mission in the concept exploration stage (involving environmental physicists and engineers from different companies). In this study, the author participated as direct observer.

Study 3 (S3) was a longitudinal study during four years with five projects in a small space systems integrator company. These projects in different lifecycle stages (concept exploration to operations) were partially consecutive and performed by a core team within which the author was thermo-mechanical subsystem officer and participant observer.

Study 4 (S4) was a four-days-study in the same concurrent design facility as in S2 but with different participants and a different project (solar science) in the concept exploration stage (involving solar physicists and engineers from different companies). Together with two interaction researchers from the DICA (Dynamics in Interaction, Communication and Activity) lab of the University of Luxembourg, the author participated in S4 as direct observer.

S1 and S2 provided first insights into interaction in space systems engineering teams and allowed for testing of data collection methods for the major studies. Formal observation protocols with pre-defined categories were considered as not sufficient, therefore observation protocols without pre-defined categorisation were performed within the two major studies (S3 + S4). Participants' effort to do participant journals (as performed in S1) was too high. Therefore, audio and video records, project journals, and email collection were identified as major data collection methods for S3 and S4. These are complemented by data collection methods of second priority research journal, documentation collection, interviews, and physical artefacts.

The data from the different sources is organized for each study into so-called study chronologies and for S3 also into project chronologies. These chronologies are databases, which support data triangulation and selective refinement of time resolution (from days to seconds), e.g. with transcripts of audio and video records, emails, project journals, and interviews.

The analytical framework comprises two methods: an activity-theoretical analysis and a theme-and-key-event analysis. The first method, the activity-theoretical analysis is based on a network of activity systems incorporating the concepts of boundary management and expansive learning (Engeström 1987). This activity systems network consists of two levels, individual and team level. An exemplary activity system of each level is shown in Fig. 2.

Six elements (the nodes) constitute activity which leads to certain outcomes. Subjects, performing an activity with certain mediating artifacts and tools, are motivated by objectives. The objectives define the activity which is governed by certain rules, situated in a wider community, and divided by different types of labour.

A zooming in on the network allows for focusing on different details, relations, and particularly on contradictions with learning potential, which are important for answering our research question. These contradictions appear within elements, between elements of activity systems, and between activity systems (on the same and on different levels). The activity-theoretical analysis includes a description of the activity systems networks and an identification of the contradictions. The contradictions, as well as identified critical interaction instances, motivate the selection of themes and key events used in the second analysis method.

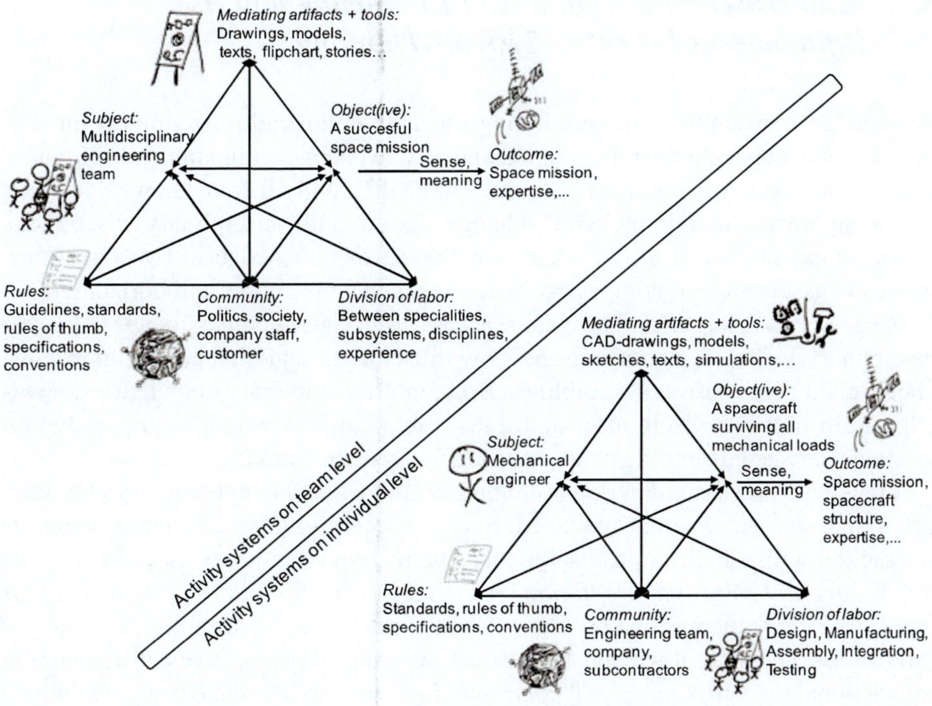

Fig. 2 Exemplary activity systems on two levels of an activity systems network

This second method, the theme-and-key-event analysis, is performed on three levels of analysis: macro, meso, and micro. The macrolevel analysis is an ethnographic description of a theme. Themes are key events that are linked in time from an emic (insider) or etic (outsider) perspective. The mesolevel analysis is performed with a dual categorisation scheme based on discourse features and systems thinking content. A systems thinking taxonomy including visualisation has been developed in order to analyse change of knowledge over time. This taxonomy, which is based on a taxonomy for learning, teaching, and assessing (Moser 2013c), allows a distinction into fields of knowledge (e.g. according to subsystems), types of knowledge (factual, conceptual, procedural, and relational), and cognitive processes (remember, understand, apply, analyse, evaluate, create). The microlevel analysis is based on multimodal interaction analysis and focuses on critical instances in interaction (Ziegler et al. 2012).

This analytic approach allows identifying contradictions with learning potential in the work activity of the four studies and factors influencing the evolution of systems thinking. A detailed description of the performed analyses is provided in Moser (2013c).

3 Findings and Discussion

3.1 *Multi-Disciplinary Quality of Interaction and Its Influence on Learning Systems Thinking*

Knowledge, relevant for systems thinking, changes within multi-disciplinary interaction. Factual, conceptual, and relational knowledge change within multi-disciplinary discussions (seconds, minutes) (Moser et al. 2012), and all four knowledge types (including procedural knowledge) change across multi-disciplinary discussions (days, years) (Moser et al. 2011). It is a change of knowledge in two directions: vertically as a change of competence within a distinct discipline, and horizontally as a change across disciplines. The change across disciplines includes changes of extra-disciplinary knowledge (outside one's own discipline) and changes of knowledge about relationship between disciplines. Both, multi-disciplinary knowledge (knowledge from multiple disciplines) and trans-disciplinary knowledge (knowledge on relations between disciplines), are required for systems thinking.

Trans-disciplinary interaction is defined as interaction where boundaries between disciplines are successfully managed. Such a successful boundary management is interaction with a high multi-disciplinary quality, which requires *awareness of the interactors' diversity and orientation towards the extra-disciplinary interactor*. This mandatory constituent is one of a set, which defines the multi-disciplinary quality of interaction. The other three constituents are diversity of perspectives, differences in interactional responsiveness, and cohesion of interaction (Moser 2013c), which will not be discussed here.

3.2 Awareness of Diversity and Orientation Towards Extra-Disciplinary Interactors

The central constituent of the multi-disciplinary quality of interaction, the awareness of diversity and orientation towards extra-disciplinary interactors, has a dual structure: the awareness of diversity and the orientation required due to the diversity. The diversity itself can be defined by different disciplines but also by different ages, languages, and cultures. Here, we focus on disciplinary differences, i.e. the diversity of disciplinary perspectives.

There are three possible combinations of "awareness of something", "no awareness of something", "orientation towards somebody", and "no orientation towards somebody". The combination of no awareness and orientation is not possible as orientation is regarded as a deliberate action, which requires awareness. The three possible combinations are explained with five examples (Example A—Example E).

Awareness and orientation towards extra-disciplinary interactors: Example A (from S4) was a theme that comprised several key events. In one key event, Sci1 (responsible for the payload of the space mission) explained the principle of umbra (full shadow) and penumbra (half shadow) four times with a different audience. The first time, Sci1 explains the principle with a sketch of a rectangular shaped occulter (shadow maker) to Sci2 (responsible for payload and mission) and Mis (mission specialist). The second time, Sci1 explains the principle with a Christmas tree-shaped occulter to a larger audience with other subsystem officers (including Str, the structural specialist). Then, the third time, Sci1 is asked by Con (configuration specialist) to repeat the Christmas tree explanation. This time he gives a step by step explanation including explicit mentioning of basic assumptions and principles. Finally, the fourth time, Sci1 and Sci2 discuss the principle in a more scientific way using different terminology. This last time is regarded as mono-disciplinary discussion, while the three earlier discussions are multi-disciplinary discussions of Sci1 with extra-disciplinary interactors.

Sci1 changes his way of explaining the principle of umbra and penumbra. He is aware of the diversity of the different interactors and selects techniques, which he considers as most suitable, such as going back to basic physics, using analogies, and natural language. The selection of a Christmas tree as an exemplary occulter shape also shows the orientation towards the interactors.

The request to repeat the explanation is an example of an expert-novice practice (Vickers 2010). The questioners position themselves as (disciplinary) novices relative to Sci1 who is considered as expert in this discipline.

In another key event, one day later, Str1 refers in a discussion with Pwr (responsible for the power subsystem) to the earlier key event (emic link) while he explains the umbra and penumbra principle. Str1, being in the novice role one day before, is now in the expert role. Such a change across discussions from questioner (novice) to explainer (expert) is an indicator of learning sustained for a certain period, here one day.

Awareness and no orientation towards extra-disciplinary interactors: Example B (from S3) shows an orientation away from extra-disciplinary interactors. Here, the diversity is defined by different languages. Switching of language was observed during meetings for the pre-launch preparation of a spacecraft. Translation interruptions expanded to parallel discussions in English, French, German, and Chinese. Not speaking English decreased the access to participants who did not understand the language in use. This can be interpreted as an orientation away from these (extra-disciplinary) interactors. Such an orientation away from interactors could be on purpose in order to hide issues. This example stresses that awareness alone is not sufficient and orientation between interactors towards each other is required to maintain a valuable interaction.

No awareness and no orientation towards extra-disciplinary interactors: Example C (from S3) shows no awareness and no orientation towards extra-disciplinary interactors. Rad1 (radiofrequency specialist) asks for dimensions within a launcher. Thm1 (thermo-mechanical specialist) mentions the height with "eight hundred" and Thm2 (thermo-mechanical specialist) takes up Thm1′s utterance, repeats "eight hundred," and completes the dimensions "sixty times sixty." Rad1 asks "sixty centimetres" and Thm1 acknowledges this with a "mhm." Rad2 (radiofrequency specialist) corrects the footprint dimension to "sixty by seventy" and Thm2 acknowledges this. Rad1 asks again for the height and Thm1 answers "eight hundred." This answer leads to the question of Rad1 if millimetre or centimetre is meant.

Soon after, it became clear that Thm1 took for granted that "eight hundred" is perceived as 800 mm since 800 cm would not make sense from his thermo-mechanical point of view. Thm2′s uptake of "eight hundred" and his completion with "sixty times sixty" (in centimetre) shows that, although Thm1′s way of interacting might have been sufficient for a discussion between the two thermo-mechanical specialists, for the extra-disciplinary interactor Rad1, this way of interacting—taking for granted that 800 must be in millimetres—is not sufficient.

This low quality part of the interaction caused clarification questions, which could have been avoided by taking less for granted and orienting towards extra-disciplinary interactors. Insufficient orientation towards interactors, i.e. taking too much for granted, is also observed in mono-disciplinary interaction, i.e. the awareness of diversity and orientation towards interactors is also relevant for mono-disciplinary interaction. In Example C, the discrepancy in understanding was identified and clarified immediately in the interaction. Two other examples show that such an immediate identification and solution of an issue is not always the case.

In Example D (from S4), interactors from different disciplines were not aware of the different perceived concepts behind certain numbers. A provided circular dimension was taken as radius by one interactor and as diameter by two other interactors. In this theme, uncovering this discrepancy took four hours.

In Example E (from S1), three interactors discussed how to study protoplanetary discs without being aware that their argumentations were based on three different perceptions of the concept protoplanetary disc. This discrepancy was uncovered when the three interactors consulted another participant after half a day. This participant asked them about their definition of the concept.

The presented examples (A-E) show that an awareness of a lack of shared knowledge is essential within interaction across boundaries between diverse contexts. Awareness of diversity is regarded as basic knowledge in other fields. Remembering an (extra-disciplinary) issue is equal to actual awareness of this issue. Orientation towards other interactors supports these (extra-disciplinary) interactors in grasping arguments from an extra-disciplinary perspective and reduces the need and the number of requests for clarification. In order to orient towards other interactors one needs to be aware of the other interactors' different perspectives, i.e. of the interactors' 'differentness'. In essence, the better the orientation towards extra-disciplinary interactors the better is the quality of interaction and the learning of systems thinking.

3.3 *Reference Repertoire as a Measure of Systems Thinking*

Interaction across boundaries, i.e. boundary management, is performed by using a set of interactional techniques such as referring to experience, physics basics, analogies, and natural language. Referring to experience was found to be the most preferred technique. In addition, referring to experience indicates past learning. Therefore, this technique is described in detail.

In literature, referring to experience while interacting with someone has been identified as a strategy of experienced designers (Ahmed et al. 2003). In our study, such references were observed in all studies, irrespective of whether the participant was a professional or a Bachelor student. References were made across projects and across lifecycle stages, e.g. within concept exploration (1st stage in space mission lifecycle) to experiences gained in detailed development (2nd stage), production and deployment (3rd stage), and operations (4th stage) of on-going or past projects (Moser 2013c). Two types of references have been identified: referring to experience in a remark and referring to experience in a narrative way. The latter one is also known as storytelling.

Referring to experience in a remark is used by interactors to display previous experience and expertise. It is used, for instance, to warn other interactors to avoid repeating a mistake, to present success as encouragement to pursue the envisaged direction, and as an indicator to show that an issue was known before (e.g. after an error or success has been repeated).

Stories are told, repeated, and spread. This includes experiences used by different interactors, in different organisations, in different situations, and for different purposes. If one re-tells a story of an experience from another team member this shows that the storyteller considers the previously told story as worthwhile to be re-told and expects this story to be meaningful to the addressed listener.

Participants referred to experiences and events which happened some minutes, hours, even years ago. The reference to such an event means that neither the learning (change of knowledge) was finished during the event to which one refers to, nor that the learning was continuously on-going until the reference was made. Principally, the knowledge could have been forgotten shortly after the event and retrieved in the situation of the reference.

To assess a change of knowledge relevant for systems thinking, the concept of *reference repertoire* is introduced. The reference repertoire includes the number and the diversity of references. Therefore, the reference repertoire changes in two ways, i.e. the number of references changes (ideally grows), and the diversity of reference changes (ideally grows). A change in the repertoire that involves references from other disciplines indicates a change of systems thinking.

3.4 A Strategy to Foster the Multi-Disciplinary Quality of Interaction

This article focusses on the first part of a larger research project. The second part focussed on the improvement of the current situation, i.e. fostering the evolution of systems thinking in practice. The underlying assumption, based on the empirical studies, was that by fostering the multi-disciplinary quality of interaction, learning of systems thinking could be improved. This improvement is the central objective of the developed WAVES (Work Activity for a Versatile Evolution of Systems engineering and thinking) strategy. One path of the WAVES strategy focuses on the continuous evolution of systems thinking. The other path focuses on the introduction of employees into entities such as tasks, teams, companies, industries, and professional life. In addition to the presentation of examples in Moser (2013a, b), a comprehensive description of the strategy is provided in Moser (2013c).

WAVES has been implemented in the company where S3 was conducted, and evaluation is on-going.

4 Conclusion

Systems thinking as a prerequisite of systems engineering, evolves in interaction within multi-disciplinary engineering teams. Relevant knowledge changes within and across discussions, in vertical (within disciplines) and horizontal (across disciplines) direction. In particular, relational knowledge, i.e. across disciplinary boundaries, is essential for systems thinking. Going beyond disciplinary boundaries is considered a key element of trans-disciplinarity in contrast to multi-disciplinarity.

The evolution of such trans-disciplinary knowledge is influenced by the multi-disciplinary quality of interaction. The awareness of diversity and orientation towards extra-disciplinary interactors was presented as main constituent of the multi-disciplinary quality of interaction. The better the awareness and orientation towards extra-disciplinary interactors is the more trans-disciplinary is the interaction between multiple disciplinary specialists. Empirical studies illustrated the importance. An example of a technique for negotiation of and going beyond disciplinary boundaries, referring to experience, has been presented.

The research project itself transcends a boundary between social and engineering sciences as knowledge from both areas is required and merged in order to develop an analytical framework. This framework, based on activity theory, enables to analyse activity and changes of multi-disciplinary engineering teams in practice, to develop a strategy to influence key factors of the evolution of systems thinking, and ultimately to implement and start the evaluation of this strategy within space industry.

Acknowledgements Thanks to all the participants of the study, the organisations LuxSpace and the Institute of Space Systems of the German Aerospace Center (DLR), Prof. Dr. Charles Max and Prof. Dr. Gudrun Ziegler, the members of DICA-lab, Prof. Dr. Lucienne Blessing, and the Engineering Design & Methodology research group at the University of Luxembourg, and to all others who contributed to this article. In addition, thanks to the National Research Fund of Luxembourg (FNR) for financing the research project within the AFR-PPP programme.

References

Adams, R.S., Mann, L., Forin, T., Jordan, S.: Cross-disciplinary practice in engineering contexts. In: International Conference on Engineering Design (ICED), pp. 343–354, Stanford (2009)

Ahmed, S., Wallace, K.M., Blessing, L.T.M.: Understanding the differences between how novice and experienced designers approach design tasks. Research in Engineering Design 14(1), 1–11 (2003)

Andreasen, M.M., Hein, L.: Integrated Product Development. IPU, Copenhagen (2000)

Blessing, L.T.M., Chakrabarti, A.: DRM, a Design Research Methodology, 1st edn. Springer London, Guildford, Surrey (2009)

Bucciarelli, L.L.: Reflective practice in engineering design. Design Studies 5(3), 185–190 (1984)

Davidz, H. L., Nightingale, D. J.: Enabling systems thinking to accelerate the development of senior systems engineers. Systems Engineering 11(1), 1–14 (2008)

Dewey, J.: How we think. Dover Publications, Mineola, New York (1997)

Engeström, Y.: Learning by expanding. An activity theoretical approach to developmental research. Orienta Konsultit Oy, Helsinki (1987)

Goodwin, C.: Seeing in depth. Social Studies of Science 25, 237–274 (1995)

Gorman, M.E.: Levels of Expertise and Trading Zones. A framework for multidisciplinary collaboration. Social Studies of Science 32(5–6), 933–938 (2002)

Hougaard, G.R.: Legitimate peripheral participation as a framework for conversation analytic work in second language learning. Forum: Qualitative Social Research 10(2, Art.4) (2009)

Kramsch, C.J.: Context and culture in language teaching. Oxford University Press, Oxford (1993)

Lamb, C.M.T.: Collaborative Systems Thinking. An exploration of the mechanisms enabling systems thinking. Doctoral dissertation, Massachussetts Institute of Technology (2009)

Long, M., Wood, C., Littleton, K., Passenger, T., Sheehy, K.: The psychology of education, 2nd edn. Routledge, New York (2011)

Moser, H.A.: Fostering the evolution of systems thinking in space systems engineering teams. In Deutsche Gesellschaft für Luft- und Raumfahrt—Lilienthal- Oberth e.V. (Ed.), 62. Deutscher Luft- und Raumfahrtkongress Stuttgart (2013a)

Moser, H.A.: Fostering the evolution of systems thinking in space industry with the WAVES strategy. In Council of European Aerospace Societies (Ed.), 4th CEAS Air & Space Conference. Linköping (2013b)

Moser, H.A.: Systems Engineering, Systems Thinking, and Learning—a Case Study in Space Industry. Doctoral dissertation, University of Luxembourg (2013c)

Moser, H.A., Max, C.J., Blessing, L.T.M.: Team learning in space projects—insights from a small satellite integrator. IAC-11-D5.2.5. In: International Astronautical Federation (ed.) 62nd International Astronautical Congress, Capetown, ZA, 3–7 October 2011 (2011)

Moser, H.A., Ziegler, G.D.S., Blessing, L.T.M., Braukhane, A.: Development of systems thinking in multi-disciplinary team interaction: two cases from space industry. In: Marjanovic, D., Storga, M., Pavkovic, N., Bojcetic, N. (eds.) 12th International Design Conference—Design, Dubrovnik, 21-24 May, 2012, pp. 1929–1940. University of Zagreb, Zagreb (2012)

Pavlov, I.P., Gantt, W.H.: Lectures on conditioned reflexes: Twenty-five years of objective study of the higher nervous activity (behaviour) of animals. Liverwright Publishing Corporation, New York (1928)

Persidis, A., Duffy, A.H.B.: Learning in engineering design. In: Yoshikawa, H., Arbab, F., Tomiyama, T. (eds.) Intelligent CAD III, pp. 251–272. Elsevier Science (1991)

Radford, J., Ireson, J., Mahon, M.: Triadic Dialogue in Oral Communication Tasks: What are the Implications for Language Learning? Language and Education 20(3), 191–210 (2006)

Schön, D.A.: The reflective practitioner. How professionals think in action. Basic Books, New York (1983)

Shadrick, S.B., Lussier, J.W.: Training complex cognitive skills: a theme-based approach to the development of battlefield skills. In: Ericsson, K.A. (ed.) Development of professional expertise. Toward measurement of expert performance and design of optimal learning environments, pp. 286–311. Cambridge University Press, Cambridge (2009)

Tan, A.R.: Service-oriented product development strategies. DTU Management Engineering, Kgs. Lyngby (2010)

Vickers, C.H.: Language competence and the construction of expert-novice in NS-NNS interaction. Journal of Pragmatics 42 (2010)

Wertz, J.R., Larson, W.J. (eds.): Space mission analysis and design. [SMAD III]. Microcosm Press, Torrance, California (1999)

Ziegler, G., Song, J.Y., Kracheel, M., Moser, H.A.: Analysing critical interaction instances in collaborative concurrent engineering: Satellite development. IJPD 17(1/2), 153–169 (2012)

The Benefits of an Enhanced Design Methodology Applied to Innovative Product Development

Philippe Blanchard, Hervé Christofol, and Simon Richir

Abstract The purpose of this study was to model an *enhanced design* methodology applied to the conception of an innovative product in a SME environment. This approach includes C-K theory in a context of disruptive innovation. In general, the industrial design process consists of four major steps: the *ego-design* phase where the designer conceptualizes a user need, a *techno-design* phase where designer and engineer find solutions to materialize the concept, a *eco-design* phase where social actors involved authorize it and then the *ergo-design* phase where the user adopts the final product. A methodological reflection leads to the modelling of the innovative *enhanced design* reasoning (where major actors are replaced by a bunch of various stakeholders). The specific SME's case was successful. Using the model, the enhanced design project management was efficient. But some more complex application cases would help secure it.

Keywords Industrial design · Innovation · Methodology · Transdisciplinarity · C-K theory

1 Introduction

The purpose of this paper is to present the proposition and the field experimentation of a model of an *enhanced design* approach. This approach includes C-K theory applied to the conception of an innovative product in an SME environment. Product or service innovations are vital to the development of most companies, especially in the context of small business industries. This *constrained* environment consists of part of the industrial population with limited resources available i.e. people, finance, and technologies.

According to studies conducted by Findeli, Carbonaro and Quarante on the evolution of design characteristics over time, we can illustrate the following relations as shown on Fig. 1 (Findeli 2005; Carbonaro 2010; Quarante 1994).

P. Blanchard (✉) · H. Christofol · S. Richir
ENSAM, Arts et Métiers ParisTech, Laval, Paris, France
e-mail: philippe.blanchard@esb-campus.fr

L. Blessing et al. (eds.), *The Future of Transdisciplinary Design*,
https://doi.org/10.1007/978-3-319-06382-9_10

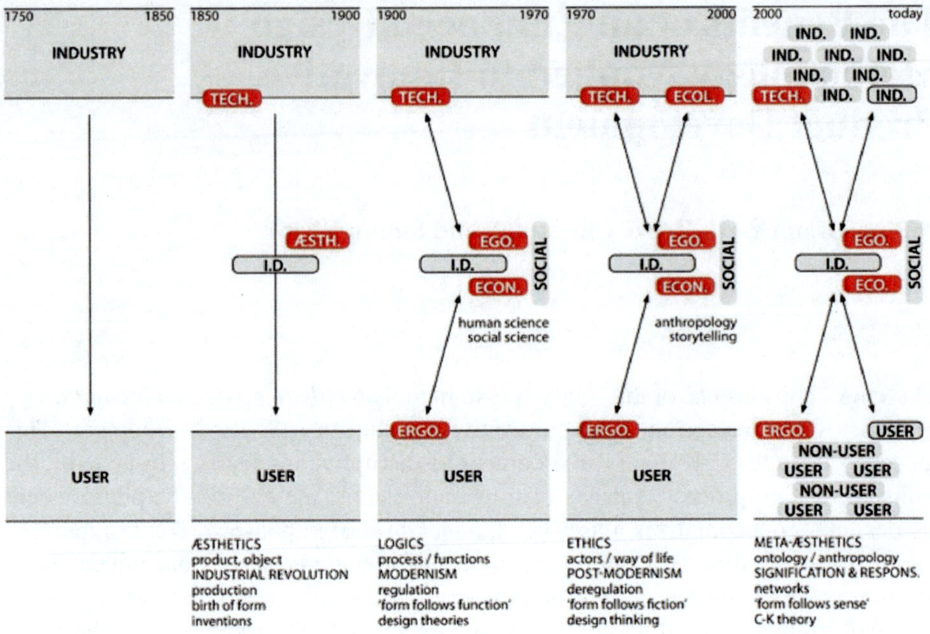

Fig. 1 Industrial Design over time

The Industrial Revolution allowed industry pioneers to provide people with manufactured goods that were obtained far quicker and cheaper than the ones previously created by craftsmen. The major challenge was the ability to industrialize the manufactured goods—in other words, produce with a machine—what was made manually before. That technical issue prevailed above anything else. The shape or form of these products was provided by technical constraints and some references to ancient famous styles. The overall production was very baroque and eclectic. Movements like Morris' Arts and Crafts demonstrated the necessity to give specific shapes to these products. As Findeli explained, an æsthetics era began where designers as form-givers had to conciliate the shape of the product with the technology of the time.

At the beginning of the twentieth century, Designers broaden their propositions. They took into account the first design theories and began to think about processes and functions— 'form follows function' as stated some followers of the Bauhaus movement. This logical approach consisted in coming up with a set of rules; a rational attitude about how to make more and more standardized mass-produced objects. Designers had to interact with technology—physical laws and machine potentialities–, with user capabilities—ergonomics, semiotics–, and with a kind of social dimension—mainly economics, hedonism. That Modernism era was characterized by a uniformity of products assigned to satisfy major needs.

In the 1970s, that products' uniformity was quite monotonous. Ettore Sottsass Jr.'s Memphis group contributed to the emergence of the Post-Modern era. It is during this period that Industrial Designers focused on a dramatic deregulation. The

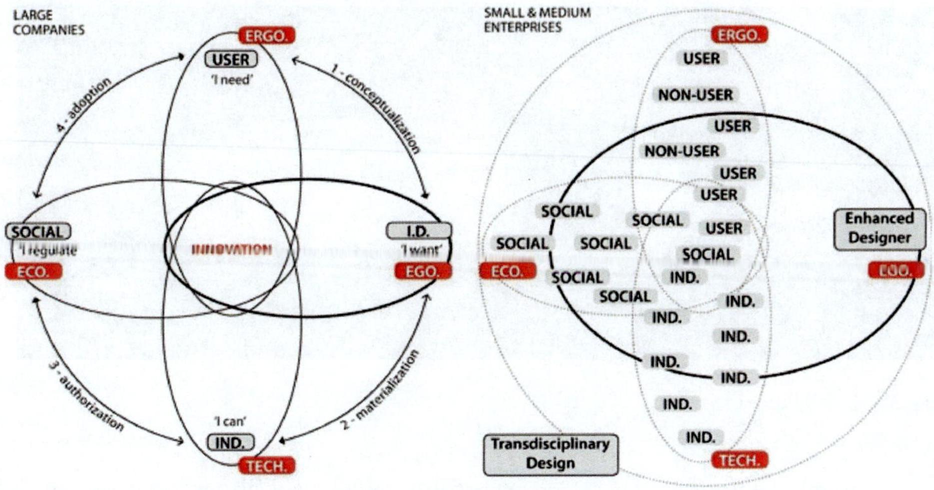

Fig. 2 The Design activity in two business environments

Design Thinking experience lead to new ways of need-finding and the imagination of storytelling objects. The latter should have a history, something to share— 'form follows fiction'. Designers were more and more involved with the stakeholders and the different experiences surrounding them. The consumer became the focal point. Their individual needs, either symbolic or immaterial were converted in products with shorter and shorter life-cycles. This resulted in an attitude of over-consumption, which was called 'excessivilization'. Confronted with this mindset of excessivilization, Designers, in an ethical perspective, tried to integrate as much as possible the new sustainable way of thinking.

At the turn of the century, the concept of signification and responsibility increased. Designers—like Ezio Manzini—were in search of meaning. They drew on ontology and anthropology as a means of giving more purpose to products— 'form follows sense'. Designers were confronted with customers that were reluctant to shopping, and in search of an authentic and sustainable existence. Their challenge was to create products or services which could bring full and nice human experiences. The business world, both global and wired, was composed of a bunch of networked entities—either on the technical side or in the user one. The Industrial Designer had to look for multiple sources of information and integrate them in his design project.

More than a century ago, Industrial Designers had to use their *techno-design* capabilities and their sensibilities in order to give appropriate shapes to manufactured objects–*ego-design.* A while after, they focused on the user acceptance through ergonomics and human science—*ergo-design.* At the end of the twentieth century, sustainable and responsive attitudes towards the environment lead to their *eco-design* approach. Each design project needed the active combination of *ego-design, ergo-design, eco-design* and *techno-design.*

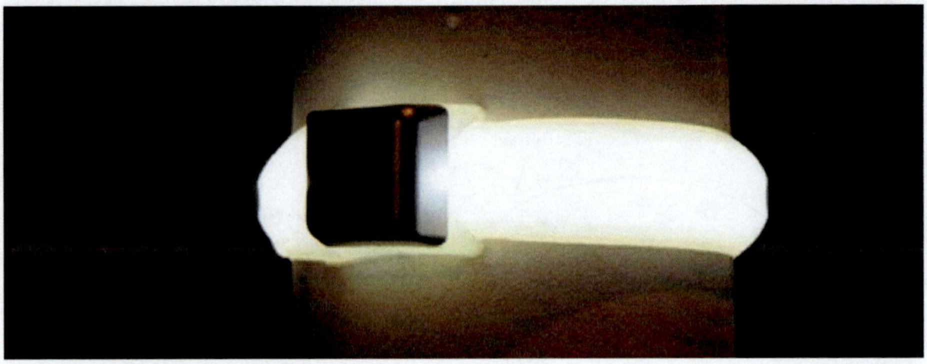

Fig. 3 The Uniklic ring

In general, the design process consists of four major steps as described on Fig. 2. At times, the user needs a product or a service, which they are sometimes unaware of. They could say 'I need…' something new, which is the *ergo-design* phase.

Hence, the Designer has to grasp this and conceptualize it in the form of a new concept, very often virtually. In that *ego-design* phase, the designer could say 'I want…' and could describe the desired future, something dedicated to the user's needs.

In the *techno-design* attitude, they could find some technical options able to materialize their new concept. The engineer could provide them with positive answers— 'I can…'.

In our social context, the *eco-design* phase is linked to both the economics and the ecology side. All actors involved—social, economic, etc. —could authorize them and then say 'I regulate…'.

Ultimately, in a last movement, the final product is proposed to the end-user for adoption in another *ergo-design* phase.

Those four different steps are not always linear and often unpredictable.

In the case of SMEs, the four actors are not very well identified. Instead of specific teams or people, the latter usually interact with various stakeholders who are mainly outside the core company. A specific team dedicated to user-centered activities could be replaced by multiple personal contacts with individual end-users or even non-users. The same disaggregation occurs in the social and the technical fields.

This is where the notion of the *enhanced designer* comes in. In general, the Industrial Designer has to work closely with technical and marketing teams. However, in an SME context, the aforementioned teams could be reduced to only one part-time person or sometimes nobody. In those situations, the *enhanced designer* has to work with their main client, the stakeholders and the final user. Designers—thanks to their cross-disciplinary skills—are able to deal efficiently with people from those specialties. Even if they do not have the total expertise of the fields, they can easily understand the main values and have the appropriate language. *Enhanced design* should therefore include 1) an artistic dimension: form, harmony, design culture, 2) a social dimension: ethnography, storytelling, ergonomics, ethics, sustainability, 3) a technical dimension: materials, natural and physical laws (Findeli 2005).

Transdisciplinary design corresponds to the subtle overlapping and synergy of all those fields. In multidisciplinarity, the different specializations are side by side with no cooperation. In crossdisciplinarity, a specialization domain is fertilized by another. In interdisciplinarity, the specialization cooperation is organized by a higher-level concept. Finally, in transdisciplinarity, the coordination is multileveled (Jantsch 1970). All the different knowledge from each domain is merged in an extended knowledge base.

The *enhanced designer* has to identify and select members of their team, from the inside or the outside of their company. As non-specialized experts, they have the common language with all those disciplines. The *enhanced designer* is able to ensure the overall work dynamism, synthesis and performance. The designer has to optimize synergy among all the stakeholders with both efficacy and efficiency. They could facilitate shared work by connecting different fields together. The *enhanced designer* is then the best person to coordinate all these fields like a conductor who organizes all the musical contributions or a director who forms his movie team in conjunction with what he imagined his project would be.

Intrinsically, a few scientific papers demonstrate the collaboration of two or three specialisms—but not all of them at the same time—and more specifically in an SME context. The selected issue encompasses the modeling of an *enhanced design* methodology applied to an innovative product development in a constrained environment. Hypothetically speaking, if we provide project leaders with the methodology it would help them enhance the performance of their new developments. It is important to note that the candidate model should be experimented and validated in the field, on a real innovative industrial product design.

Contrary to Scientists, Designers work more in applied research; they take 'use-inspired' principles and they develop applications for them (Driver 2011). Designers deal with an increasing numbers of areas. They constantly need to better their transdisciplinary skills.

In today's globalized world, everything is evolving faster. Innovation is more intense–more and more products are being designed. New solutions are being sought–surprise or breakthrough propositions. Innovation involves group work i.e. moving away from the idea of an inventor working all on his or her own (Garel 2012). The need for breakthrough design or disruptive proposition implies finding something very different from the common archetype—also called 'dominant design', the representation anybody has. The collective aspect has a link with the open-innovation. Here people can either have access to new technology not proprietary or can provide their own know-how to others.

2 Approach

2.1 Building the First Models

A methodological reflection leads to the modeling of this innovative enhanced design activity. Over time, three major means were used to model design reasoning. Initially, researchers talked about 'systematic design' (Agogué 2013). Then, designers began to define the 'design thinking' approach (Amaral 2011). Finally, the most recent developments of Hatchuel's team deepened the C-K theory as an effective tool for an innovative design (Hatchuel 2009). A summary of all the different approaches will be explained below.

Technical design models. At the beginning of the nineteenth century, the term of engineer appeared. Some tried to analyze their reasoning method used during the development of new industrial products. A first prescriptive approach by Hubka and Eder, improved by the reflections of Pahl and Beitz, lead to the definition of the systematic design mechanisms (Hubka 1987; Pahl 1996). It consisted of the application of four different phases. For the first one, the project began with the clarification of the task—finding the associate knowledge, framing needs, planning the future actions. The second one corresponded to the conceptual design where candidate concepts merged—exploration of variant propositions, experience of diverse combinations. The third one was the embodiment design where the selected concept took shape—adjusting, materialization, refinement. Then the last one dealt with detail design; the selected solution was clear enough for a transmission to industrialization specialists. The entire process was too often perceived as linear. However, the rigorous separation into different phases helped the management of complex projects and a great variety of industrial products. Later on, Simon took into account the cognitive and human variability of each one involved in the conception's activity (Simon 1996). He described the mechanism of reasoning with the analogy of the personal computer—a brain-processor that handled and studied some data-memories as mental representations. Schön insists on the attentive observation of the designer during and after his design (Schön 1983). The designer was a hypothesis provider and each of them has to be experimented and evaluated. They have to observe everything surrounding the subject—*what existed*–, to imagine and build mental representations of the problem, then distort them until they obtain a result that makes sense—*what could be done.*

Henceforth, one notices that during the design process the two ends of the process–the problem and the solution—each evolving separately in their respective areas. The designer tries to reframe a wicked problem by creating original mental representations, some specific translations of particular situations or through adventures in real new territories. In this way, they can imagine a 'mental prototype' that is very useful for the understanding of the initial problem. Then, with some loop experiments and some breakdown into sub-problems to solve one by one, their process will lead them to an imagined, tested and validated solution (Cross 2001). Gero proposed the FBS model—Function Behavior Structure—that shows the various interactions between a desired space and the real world (Gero 2004).

To sum up all the different research, Choulier suggests a generic scheme where input and output are well defined; as well as the elementary sub-problem division, each of them had to be solved one by one with iterative loops (Choulier 2008). When all the aspects of the initial issues are identified, solved and integrated, a solution to the design issue is given to the industrialization experts.

'Design thinking' models. Initially, Simon could be considered as a pioneer of theorist reflection. Then, the focus is put on collaborative techniques. IDEO did a lot for the emergence, the experimentation and the diffusion of the 'design thinking' approach: books from Kelley and Brown contributed to this (Kelley 2002; Brown 2009). The process could be described in three different steps. The first one is the 'inspiration' where people had to immerge themselves in the world and give the issue a new formulation. The second one called 'ideation' is where the designer has to be creative and find original solutions. The last one is the 'implementation' where the selected answer has to be put in order and then begin again for another loop cycle (Beckman 2011).

C-K theory principles. Disruptive innovation occurs when the proposed object is totally different from its 'dominant design'—the way it is commonly perceived; the archetype, the usual answer, the first image going in mind. The innovative designer's challenge is to enhance an actual situation with a tangible new proposition. The gap between today's products and a future solution is rather difficult to cross. There is a kind of genealogy of objects with specific lines; and those lines shaped a common identity reference. That image is quite present in each designer's mind. It is truly important to defix it in order to free the path to new possibilities both about shapes or functions. The short period of time without any tangible reference is a difficult one for design students. Traditional design theories are not so well adapted for disruptive situations—in case of technology breakthroughs for example. They are most efficient when the studied object is quite well identified (Agogué 2013). A new design theory should take into account the new identity of disruptive objects. At l'École des Mines de Paris, Hatchuel, Weil and Le Masson defined, experimented and spread the C-K Theory (Hatchuel 2009). It is both a design theory and a theory dealing with the mindset used during the conception.

The model is structured in two distinct spaces: the first one—K space—gathers the knowledge and the second one—C space—deals with concepts. In the K space, all propositions have a logic status; people can determine if they are true or false. Whereas in the C space, propositions have no logic status: no one could determine whether they are true or false, they are 'undecidable'. Designing with this theory consists of starting from an initial undecidable root concept—C_0. Then a double expansion, both in C space and in K space, with crossed operators from C to K and K to C, will enrich the root concept C_0 in order to describe it sufficiently for a K validation. All those interactions are drawn in a C-K diagram that shows the reasoning path. The C expansions occur when a K attribute is added to or subtracted from the C_n studied concept.

The model is useful to get rid of strong identity products. Also, the double expansion allows the crossing of concepts and knowledge, which prevent the validation of a good idea but absolutely unrealistic or inapplicable.

Synthesis proposition. To sum up, some notions are very important to keep in mind: co-evolution, immersion, mental prototype, sub-problem division and one by one problem solving, multiple interactions—within and outside the design team.

After the constitution of a project team, the generic process consists of a cycle of different steps. At first, the enhanced designer has to immerge himself in the problem—or situation—context. He or she has to interact with all the different domains listed above—industry and technologies, user experience and social issues. This inspiration phase associated with an attitude of empathy will help him or her form a K base. From the entire gathering, some images are likely to emerge. Many attempts to synthetize or to try some new formulations will help the imagination of a mental prototype. The ideal and desired sketch should orient and drain a flow of ideas to refine and test. The mental prototype could be viewed as the C_0 from the C-K theory. It is a root concept, an undecidable objective but still has a lot of potential and it will attract future propositions—in C space.

The sub-problems division shows the progress of the concept's expansion. The central model place is where all decisions are made. To answer an identified sub-problem, any candidate proposition is analyzed there. According to the team's desired criteria, the test is carried out and a decision follows. If it is validated, then the studied sub-problem is solved and the next one is immediately activated. If the expected characteristics are not met, the proposition is rejected and a new one has to be found for the same sub-problem. If no solution is found, then there is a need to go back to the previous sub-problem division and imagine a new one. Those back and forth movements imply the co-evolution of both the problem and the solution.

These notions were incorporated into an enhanced design process model with a specific symbolic representation. In order to build it, it was confronted a posteriori with the reasoning process used for some successful design studies. From that experience, four different activities were identified according to four different axes in the model. The first axis corresponds to *ego-design*, the shape, the personality, and the specific contribution of the designer—as a form giver. The second axis called *ergo-design* deals with ergonomics, usage, functionalities; it concerns the designer's skills—with added marketing and engineering. The third axis takes into account the *eco-design* both economy and ecology. It lists responsive and ethics criteria, with the help of the marketing field. Finally, the fourth axis named *techno-design* refers to engineering, tangible producing and operating aspects, everything relevant to the engineering expertise.

In the representation of that enhanced design process model (Fig. 4), some mini C-K diagrams were used to explain the kind of mental reasoning used during the innovative design process. The diagrams, symbolized by little capsules, showed the respective K space and C space as well as all the interrelations between them. The first capsule named 'needs' represents the immersion phase and the C_0—or mental prototype—proposition that would drain all future propositions. The central diamond is like a 'processing table' where all the sub-problems and the associate solutions had

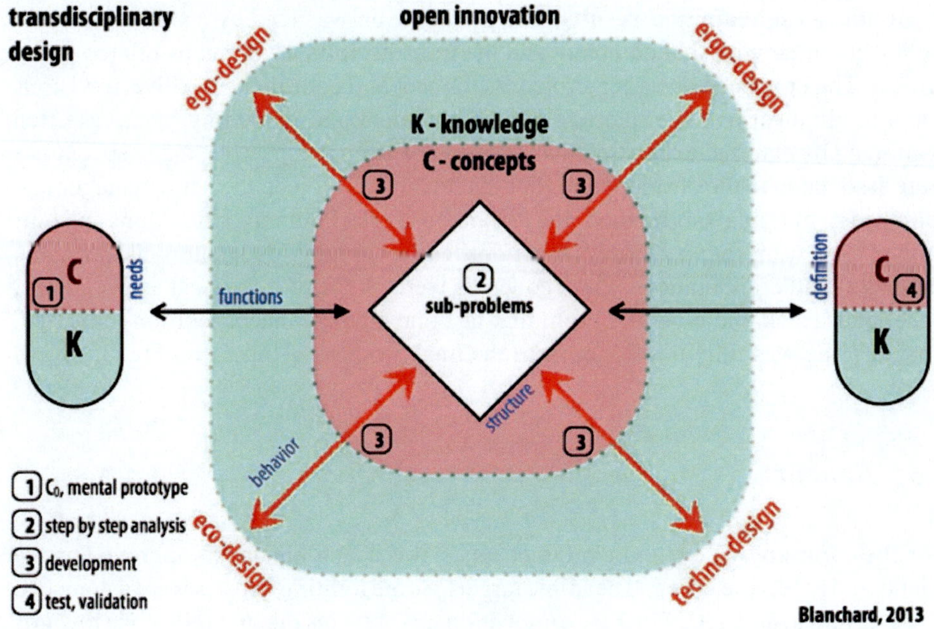

Fig. 4 The enhanced design process in progress

to be analyzed one by one. The current process *needs, function, behavior, structure, definition* is well integrated (Perrin 2001). When all sub-problems are solved, the candidate proposition is validated according to each criterion—or axes–. Then, its definition is sufficient and it can be moved from the C space to the K space, and the product development will go on with the industrialization phase with the engineering team.

2.2 Experimental Context

The enhanced design process model had to be tested in a real product development project in order to be confirmed.

SME choice. *TMC Innovation* is a small company of almost twenty people. Its mission is to improve the public area with an appropriate lighting. Faced with cost reduction, some cities turned off the lights in the middle of the night and the sidewalks became unsafe. In that specific case, the company developed a signpost solution instead of the traditional lighting system. A LED strip fixed to the pole, during the production process, uses only 1 W/h, in comparison with a 100 W/h lamp's consumption. *TMC Innovation*'s clients were so enthusiastic about this device that they asked to implement it on already existing poles. There is a high demand for it and the variety of poles' geometry make that adaptation rather difficult. That specific subject was chosen for the enhanced design process model experimentation.

Building the team and results. The project team was lead by a skilled designer well acquainted with that company and the transdisciplinary domains related to the project. The enhanced designer's role was to meet and coordinate the diverse visions about the innovation. The internal team, a technical and marketing one, was often reinforced by external expertise. In an open-innovation perspective, many specialists were associated to the development. For that technical subject, the team management followed a value analysis methodology (Yannou 2004). Four major functions and four limited ones were found. Those 8 functions lead to 30 individual solutions. After the validation and combinations, three concepts were chosen for the next step: 'donut', 'lace' and 'stackable objects'. From that last one a new concept 'cordon' emerged. That last one was fully developed until the final product—Uniklic on Fig. 3.

2.3 *Benefits*

For the company. The Uniklic product exists; it is a tangible one, is approved by the clients and meets the cost and deadline targets. Some lighting experts were astonished by its technical audacity. For *TMC Innovation* it represents the first milestone towards an innovative new products strategy.

For the modeling. The enhanced design process model is strengthened by a successful real size experiment. The first hypothesis, the interest to model the design process and apply it to an SME context, is reinforced. Thanks to that model, the project management was efficient, both inside and outside the design team. The major steps of that development are shown in the model—Fig. 4. Initially, the C_0 or mental prototype was 'how to fix a LED strip onto a lighting pole'. Secondly, with the value analysis method the problem was framed and divided into many sub-problems to be solved one by one with concepts and candidate solutions. Then, the field validation transferred the Uniklic *cordon* from the C space to the K space. Finally, the specific development resulted in the addition of the new product in the company's catalogue.

3 Conclusion

Often, due to limited resources available, SMEs cannot afford to recruit multi-specialized teams. So, each team member has to be extremely versatile and take into account various aspects of the design development process. From a transdisciplinary perspective, the enhanced designer has to interact with the marketing and technical teams in order to expand the initial industrial design territory. Going back and forth according to the principles of open-innovation would complete these three skills. Using this model would be a success factor. Some more complex and less defined application cases—other than the Uniklic one—would need to be explored. This would help secure the model.

References

Agogué, M., Arnoux, F., Brown, I., Hooge, S.: Introduction à la conception innovante, éléments théoriques et pratiques de la théorie C-K. Presses des Mines—Transvalor, Paris (2013)

Amaral, C.: A Design Thinking Approach in Startups. Macromedia Hochschule für Medien und Kommunikation, Munich (2011)

Beckman, S. L., Barry, M.: Innovation as a Learning Process: Embedding Design Thinking. California Management Review, 50, 25–56 (2011)

Brown, T.: Change by Design: How Design Thinking Transforms Organizations and Inspires Innovation. HarperCollins Publishers, New-York (2009)

Carbonaro. S. Votava, C.: Form Follows Sense. New Innovation and Design Strategies for Crisis-Ridden Times in *Creating Desired Futures.* Birkhäuser, Basel (2010)

Choulier, D.: Comprendre l'activité de conception. *Chantiers* de l'Université de Belfort-Montbéliard, Belfort-Montbéliard (2008)

Cross, N.: Design cognition: results from protocol and other empirical studies of design activity. Georgia Institute of Technology, Atlanta (2001)

Driver, A., Peralta, C., Moultrie, J.: Exploring How Industrial Designers Can Contribute to Scientific Research. International Journal of Design, Vol. 5, N°1 (2011)

Findeli, A., Bousbaci, R.: L'Éclipse de l'Objet dans les theories du Projet en Design. The Design Journal, Vol. 8, Issue 3 (2005)

Garel, G., Mock, E.: La fabrique de l'innovation. Dunod, Paris (2012)

Gero, J.S., Kannengiesser, U.: The situated function behavior framework. Design Studies (2004)

Hatchuel, A., Weil, B.: C-K design theory: an advanced formulation. Research in Engineering Design 19:181–192. Springer, Heidelberg (2009)

Hubka, V., Eder, W.E.: A scientific approach to engineering design. Design Studies (1987)

Jantsch, E.: Inter- and Transdisciplinary University: A Systems Approach to Education and Innovation. Policy Sciences 1 403–428 (1970)

Kelley, T.: The Art of Innovation: Success through Innovation. Currency Book Doubleday Random House, Inc., New-York (2002)

Le Masson, P., Hatchuel, A., Weil, B.: Strategic Management of Innovation and Design. Cambridge University Press (2010)

Noailles, P.: De l'innovation à l'innovateur, pour une approche structuraliste de l'innovation. Revue des Sciences de Gestion. Paris (2011)

Pahl, G., Beitz, W.: Engineering design, a systematic approach. Springer, London (1996)

Perrin, J.: Concevoir l'innovation industreielle: méthodologie de conception de l'innovation. CNRS Éditions, Paris (2001)

Quarante, D.: Élements de design industriel. Polytechnica, Paris (1994)

Schön, D. A.: The reflexive practitioner: how professionals think in action. Basic Books, New-York (1983)

Simon, H.: The sciences of the artificial. MIT Press, Cambridge MA (1996)

Yannou, B., Bigand, M.: A curriculum of value creation and management in engineering. European Journal of Engineering Education, Vol. 29, Issue 3 (2004)

Transdisciplinary Research—Buildings as Service-Oriented Product-Service Systems

Cyril Mauger, Jean-Yves Dantan, Ali Siadat, Eric Dubois, and Sylvain Kubicki

Abstract This paper proposes a *transdisciplinary* research for the conceptual definition of a building. It is based on the concept of Service-oriented Product-Service System (PSS) applied to buildings. Existing kinds of PSS are about service(s) integration whereas Service-Oriented PSS is about product(s) integration. The product definition and design aims to be well-aligned with the services the customer wants to provide to his community and the chosen means to deliver them. The main focus is on the services and how the building could support it or contribute to its improvement. As a consequence, the building as well as the services becomes parts of a "building system" to be defined with the customer. This enlarged way of seeing a building required the use of a *transdisciplinary* research. It provides new perspectives of innovation in its definition and a better integration/alignment between all of its components during the design phase. It calls for more accuracy and methodology in the service definition in order to ensure this alignment.

Keywords Integrated product service · Conceptual design · Product-service alignment · Construction · Architectural programming

1 Introduction

Transdisciplinarity—this concept denotes a sublation (i.e. transcendence, from the German term Aufhebung) of disciplinary boundaries. *Transdisciplinarity*, as opposed to *multidisciplinary* and *interdisciplinary* (Jantsch 1972), concerns that which is simultaneously between disciplines, across different disciplines, and beyond disciplines (Nicolescu 2005). Its goal is to understand the present world in a more systematic and holistic manner, of which one imperative is the unity of knowledge Max-Neef (2005).

C. Mauger (✉) · E. Dubois · S. Kubicki
Luxembourg Institute of Science and Technology, Esch-sur-Alzette, Luxembourg
e-mail: cyril.mauger@m4am.net

C. Mauger · J.-Y. Dantan · A. Siadat
Laboratoire Conception Fabrication Commande, Arts et Métiers ParisTech, Metz, France

L. Blessing et al. (eds.), *The Future of Transdisciplinary Design*,
https://doi.org/10.1007/978-3-319-06382-9_11

Our research work is based on Wiesmann et al.'s definition of transdisciplinary research (Wiesmann et al. 2008) (proposition 1): "*Transdisciplinary* research is research that includes cooperation within the scientific community and a debate between research and the society at large. *Transdisciplinary* research therefore transgresses boundaries between scientific disciplines and between science and other societal fields and includes deliberation about facts, practices and values." Our aim is to propose a transdisciplinarity framework for building requirements definition.

Construction is already considered as a complex system (Bertelsen 2003) especially in its construction phase (Kubicki et al. 2006). The Built Environment is already accepted as a *multidisciplinary* field (Kalay 2001). For a while, it has been influenced by Management discipline developing a strong Management-led *crossdisciplinarity* and now it called for *interdisciplinarity* (Chynoweth 2006). Urban studies have already been subjected to *transdisciplinarity* on research experiences in France and Canada (Ramadier 2004).

The research in this paper goes further in this direction by proposing a *transdisciplinary* research (i.e. through the following disciplines: Mechanical engineering, Industrial engineering, Software engineering, Construction engineering) about another phase of the Built Environment: architectural programming (i.e. requirements definition part of buildings' conceptual design phase). One of the main results leads to a *transdisciplinary* requirements definition approach. This approach is needed due to our two original propositions about buildings definition. This first one presents buildings as Product-Service System (PSS). It calls for the correct alignment between the service, the building, and its components. The second proposition introduces a new kind of PSS mindset that better fits to buildings: Service-Oriented Product-Service System. The requirements definition of building is therefore led by the definition of the services to provide inside the building. The building is thereafter considered as a support for the service that achieves the business.

The aim of this paper is to justify the development of a *transdisciplinary* approach for architectural programming: Why a sublation of the disciplinary boundaries? We propose a response by defining the scope of the studied system. Therefore, the Sect. 2 details the scope of the building system as a PSS; the Sect. 3 focuses on its characterization as a Service-Oriented Product Service System. Finally, the last section introduces a discussion about the designed *transdisciplinary* approach.

2 Buildings as PSS Mindset

A building is a multilayered complex artifact. Technical building services introduce its first level of complexity in terms of design and construction project management (Kubicki et al. 2006). This paper focuses on a second level of complexity related to the main purpose of a building in a preliminary phase: the requirements definition/architectural programming. In public construction, this purpose is to shelter and support a human intensive activity we call Service. This service is just outlined and not so often revised regarding standards during the architectural programming. As a result, the requirements specification tends to be focused on description of technical solutions related to the building with little integration of the service to provide.

Despite the differences claimed by construction professionals about each project, they do not fully exploit one of the core origins of these differences: the local context and its major impact on the services to provide to the local community. Each project has a specific social, economic, politic context. Needs are thereafter very distinct and require a tailored solution. This solution depends upon the people, resources, and organization that make up the services rather than upon the building itself. A secondary school in the suburbs of Paris does not have the same education goals and cultural environment or answers the same education and workforce issues as a secondary school in Luxembourg even if the building is quite the same. The definition of their building should be enlarged to integrate their specific services instead of following standard patterns of definition leading to copy-paste solutions.

In order to explain why buildings could be seen as PSS, let's take a concrete example based on a case study partially handled by the research team: the requirements definition of a new public school. The main service provided by a school is education. The owner of the building is the same as the owner of the service i.e. the State. The State imposed directives to the superintendent and asked him to apply them in the future school. According to these directives and in association with his board and sample of teachers, the superintendent chose a new leitmotiv associated to this school: "Learn to be". They wanted to encourage mutual aid, communication and team building. This leitmotiv would lead reflections about the future building.

One of the main issues with the former building concerned bully zones. After an analysis of the previous design (Fig. 1), it appeared that the only "safe zone" in the school was situated near the administrative and teachers area. Regarding the leitmotiv of the future school, things had to change. The research team guided the contracting owner and his task force on this issue amongst others. First step was to define the proper requirement on the system (i.e. the building as a black box) using a Functional Analysis approach. The job was not easy because the task force had a solution

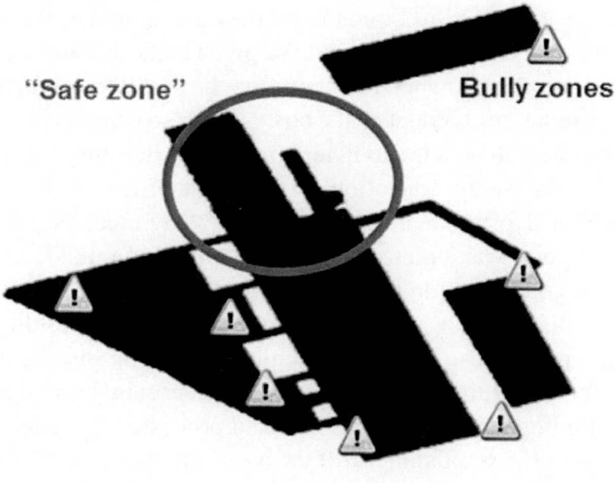

Fig. 1 Analysis of the former building plan view from above

thinking process instead of a requirement-oriented mindset. The corresponding high level requirement formulated was that every pupil inside the school could be seen by an adult. The second step was the generation of solution principles (i.e. programmatic concepts, different from design solution). The main difficulty to overcome by the research team consisted in making the task force understand that they were not limited to single principle such as the building or additional expensive equipment.

Several solution principles were generated and retained by the superintendent's team to fulfill this main requirement:

- *Architecture and positioning.* The architect could propose a different layout and design shape of the spaces for watching improvement (like in prisons). Requirements to provide were about proximity between particular spaces.
- *Transparent materials.* The architect could favor the use of see-through materials to prevent bullying by lack of spaces privacy. Requirements to provide were about kind of materials to be used and visibility to and from particular spaces.
- *Teachers and rounds.* The superintendent could ask school's personnel to do some rounds scheduled by pair during the breaks. Independently from the design of the building, all the requirements concerned only the organization and people activity.
- *CCTV system and guards.* Independently from the design by the architect, the superintendent could install a CCTV system and mobilize the janitor for supervision. Requirements would concern equipment, network and the janitor's office.
- *Hybrid solution principle.* The last possibility would be to combine positioning of the adults space around the school with a direct view on common spaces and circulation, the use of see-through material for their offices in order to allow a passive watch of the pupils by the adults. Requirements to provide would concern proximity of spaces and materials used for the walls.

Each one of these solution principles had an impact on different parts of the school: e.g. the building, the organization, the people or the equipment. Requirements on the building then differed and would give orientations toward the proper solution. The role of the architect is to design buildings, not organizations. The choice between alternatives had to be made by the customer (i.e. the superintendent). Then, correspondent requirements would be given to the architect. It was part of his responsibility as contracting owner. Furthermore, the architect could not make this decision instead of the superintendent as the cost of each solution principle would not only impact the building but also the long term use of the building, i.e. the organization. His decision was lead by the education service to be provided. It did not aim at watching pupils' acts or deprive them from privacy by any means as in a maximum-security prison. The point was on ensuring their well-being outside class hours to avoid stress and poor concentration due to bullying and intimidation.

This example presented different parts of a building by exploring a set of solution principles. The last principle focused on the building as a way of ensuring the safety of pupils with as little impact as possible on other components of the building including day-to-day activities and processes. It underlined the fact that the building was not a standalone artifact but a component of a bigger system to be defined

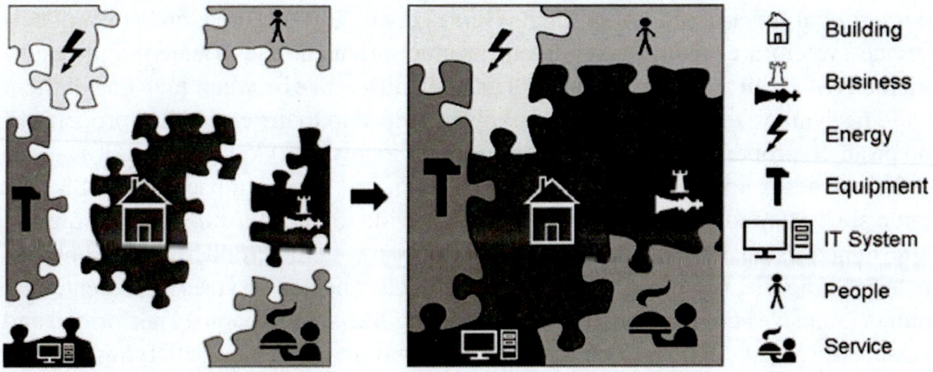

Fig. 2 Composition of the "Building System" as an aligned PSS

according to the service to be provided. As a consequence, the building as well as the services becomes a part of a "building system" to be defined by the customer. Other parts of this "building system" are: the people who work inside it, the equipment required to support the people's work, the energy to be supplied to the equipment and building, the information technology system which stores information about people and processes, and the business which organizes and manages the whole system. Definition of each part should be done during the requirements definition phase in an integrated process to ensure consistency and alignment between them (Fig. 2).

Current practices in construction leave the customer focus on the building with few in depth discussions on the other parts and no proper formalization of the solution principles. Interactions between the building systems' parts, their synergies, are not developed as it could be. They are defined later depending on the design of the building and most of the time after construction of the building. As a consequence, a couple of patches are required to adapt the building leading to extra costs and delays. PSS mindset could contribute to improve the degree of integration and alignment of each components of a building system. Each one of them is related to a specific discipline.

3 Service-Oriented PSS

Product-Service System (PSS) is mainly presented in literature as a business model. It focuses on functionality or usages to provide to consumers instead of selling products (Meier et al. 2010). The idea is to sell a marketable mix of products and services that will jointly satisfy the consumers' needs (Goedkoop et al. 1999) and increase at the same time the market proposition (Mont 2000) by integrating services to traditional functionality of products Baines et al. (2007). Three kinds of product-service system are mainly considered: product-oriented, use-oriented, and result-oriented (Manzini and Vezzoli 2003). The main difference between each one of them could be synthesized into the degree of ownership sold to the consumer: property of the product, property of its use, and property of its results (Cook et al. 2006).

This paper is not about business models, even if it has an impact on it. PSS is at first a shift of mindset about the solution to provide to the consumers. This solution is designed as an integrated product and service system. Products are defined as material, tangible, produced, and storable artifacts whereas services, represented by human intensive activities, are immaterial, intangible, simultaneously performed and consumed (Moritz 2005). Each part of the system is capable of satisfying specific consumers' needs, complementarily or jointly (Manzini and Vezzoli 2003), through an anticipated integration from the early stages of development. A proper alignment between each part of the PSS provides synergies and added values for providers and consumers in a win–win situation. PSS are expected to significantly reduce resources deployment and improve their efficiency (Cook et al. 2006). This mindset is setup in the earliest phase of a project, during the requirements specification and conceptual design phase.

We proposed in this section to define a different kind of PSS requirements definition mindset that would fit with the proposed view of the "building system": the concept of Service-Oriented PSS. Focus was on the requirements definition to ensure a full integration of services in the specification of buildings and their components. The Service-Oriented PSS was defined regarding an interpretation of the other three kinds of PSS as mindsets for requirements specification and design.

The main interest of PSS resides in the integration of services at different levels of products development. Literature considers three main kinds of PSS Manzini and Vezzoli (2003). The starting point is always a product and the end point services, associated or substituted (represented by arrows in Table 1).

- *Product-oriented PSS* starts from a product and adds services as a support of its use along its lifecycle. The product is developed so as to derive services controlled by the provider such as warranty or maintenance (e.g. Apple through its own standard connectors). The product and the services can be seen separately for the customer.
- *Use-oriented PSS* transforms the product into a more flexible leasing service. The product is owned by the provider but used by the customer (e.g. car sharing). The

Table 1 Shift of paradigm toward PSS

	Product (P) / Service (S)			
	Start point	**End point**		
Product-Oriented PSS	P	→ P	+	SSS
Use-Oriented PSS	P	——→	PS	
Result-Oriented PSS	P	———	——→	S
Service-Oriented PSS	S	→ PPP	+	S

product is visible through the service by the customer. Usually, expected qualities of the service have to be negotiated with the customer.

- *Result-oriented PSS* shifts from selling products to providing services. The product is owned and used by the provider (e.g. industrial washing machines). It allows designing more complex products reflecting the providers' expertise and guarantying a higher efficiency in their use. The product is invisible or disappears from the customer sight. It lets place only to a final result. Usually, service contracts between the provider and the customer are formalizing the expected service qualities level.

In each case, the design of the product is impacted by the services associated. The same product can evolve toward each one of these models. As a result, three different designs could be proposed focusing on different points (e.g. specific spare parts, high resilience of the product, or efficiency of the product).

A *Service-Oriented PSS* starts from an existing service accompanied with a set of products used as supports (Table 1). Usually, these products are designed independently from the service or with few integration of it in the development phase. This kind of services requires "tailored" product during their provision such as cars (i.e. cabs) for taxi drivers or, in this case, school buildings for public education. Integration of the product to the definition of the service leads to improvements through a better alignment. The service is the main focus but the product gives it consistence. The product's design could have an important impact on the service delivery for this kind of PSS. This impact can be either good or bad, and service is quite sensitive to it. Poor design of the product leads to poor service delivery. This poor design is caused by a poor understanding of the service activities and a lack of comprehensiveness in its processes description. Function or name of the product tightly depends of the service. The service is the main part of its identity. The design of the support products depend on the service to be provided and not on the business model as for the other kinds of PSS. Existing kinds of PSS are about service(s) integration whereas Service-Oriented PSS is about product(s) integration. The idea could be compared to Shimomura's description of the essence of service design (i.e. "what to offer" and "how to satisfy customers" (Shimomura et al. 2009) as a first step toward products definition/integration. Depending on the service to provide, derived functions could be implemented by humanware or hardware, i.e. the different kinds of components of the PSS. This description of Service-Oriented PSS fits with the view of buildings proposed in Sect. 2. From the same objective to achieve, different requirements could be defined and not only about the building for solution principle.

4 Discussion

In this paper, buildings are first presented as PSS then characterized as Service-Oriented PSS. This section presents why *transdisciplinarity* was introduced in the research framework to support this new paradigm.

Table 2 How disciplines deals with Product and Service: an empirical assessment

Discipline	Product		Service
Mechanical Engineering	+++		Ø
System Engineering	++	←	+
Software Engineering	Ø		+++
Enterprise Architecture	Ø		+++
Industrial Engineering	+	→	++
Architecture	+++		Ø
Proposal	+++	↔	+++

The disciplines gathered in this research work are Mechanical Engineering (ME), Software Engineering (SE), Industrial Engineering (IE), and Architecture. Each one of them mainly focuses either on products or services (Table 2):

- In ME, most design approaches are focused on product solution research based on user requirements. These are inferred by careful analysis of similar products. Existing solutions are reconsidered but not their use or related service. System engineering design approaches Kossiakoff et al. 2011) deal with complex mechanical systems but are mainly focused on physical products rather than technical services associated or added to the complex system Maussang 2008).
- In SE, central IT products are software delivering functionalities to their end-users. A Product-oriented PSS has been considered from the beginning with associated services like e.g. maintenance. Then, the Software as a Service (SaaS) model has been introduced in the early 2000 Turner et al. 2003). It privileges a User-oriented PSS where the ownership of the software is abandoned. More recently, the software itself is becoming 'invisible' (no information about its location, neither its ownership) with the recent cloud model associated with a Result-oriented PSS. Finally, Service-oriented PSS is now emerging Demirkan et al. 2008) with the more holistic service system perspective where the software is just a component from a larger system defined by Spohrer Spohrer et al. (2007) as "a configuration of people, processes, IT and shared information connected through a value proposition with the aim of a dynamic co-creation of value through the participation in the exchanges with customers and external/internal service systems".
- In IE, the main goal is to propose a set of general methods and tools for product and service development. It takes into account that specifications of products or services should be developed but does not take advantage of relationships between these two concepts. Usually, a service is considered like a product. PSS design in IE tends to be Product-oriented whereas customers seek for Result-oriented PSS. Hussain et al. Hussain et al. 2012) proposes a framework to support this shift of mind-set using system-in-use data.
- In architectural design processes, architects contribute to the design and construction of building facilities (product). Tools are developed to enable the modeling and simulation of this built environment as well as to improve the multi-expertise design processes themselves Eastman et al. 2011). In most of the current design

projects the service developed within the (future) facility is not formally defined, but simply outlined in the brief report. Therefore, interpreting it remains the mission of the designers (architects, engineers) but usually lacks rigor and traceability.

- In this paper, the building is no longer considered as a standalone product but as a Product-Service System. Its requirements definition called for transdisciplinarity to properly and equally integrate and challenge its product part and its service part.

In Sect. 3, a new mindset for the requirements definition of buildings is proposed. The service to provide determines the requirements about the necessary resources including the required building through the Service-Oriented PSS mindset. As a result, the requirements definition framework of the system starts with the higher level of abstraction, regarding the service to provide, goes through all its components and ends with the building which hosts everything (Table 3).

Using this second level of details, each discipline was analyzed regarding the components of the building system. Table 4 refines the empirical assessment of each discipline about their focus (+ + +), integration (+ +) or impact (+). The relationships between the artifacts in each discipline are at the center for the design of this proposal.

The requirements definition of such a building system cannot thereafter be done using current approaches. It gathers too many concepts/artifacts in the same system. Specific discipline partly addresses some of these concepts but not the whole paradigm. Architecture does not fully support it either. Regarding these observations, a *transdisciplinary* research is required to properly address the definition and design of future buildings. ME brings structural decomposition and "product" part concepts. System Engineering completes it with the integration of "IT" part concepts. SE improves this IT integration making the links with the "service" part concepts and humanware. Enterprise Architecture deepens the service integration by adding a "business" layer. IE enhances the relationship between "service" parts and material resources. Architecture finally brings the last piece: the "building" part which hosts the whole defined system.

Core part of this transdisciplinary research lays on the principle of "bridging" disciplines through their similarities to benefit from their specificities. Each discipline provides specific viewpoints on the studied system independent from each other. These viewpoints refer to each sub-system defined by their own vocabulary, described by their own models through a requirements definition performed following their own methods. Nevertheless, the concepts used for definition (i.e. meaning, sense of words), the questions answered in the process (i.e. what, why, who, when, where, and how), and the logic followed by requirements engineers and designers are the same. A semantic issue between disciplines has to be overcome to consider interoperability between their respective techniques, models and methods. The main difference is the nature of the outcome, specific to their domain. The scope of the studied phenomenon (i.e. the building) was enlarged to integrate elements of its immediate environment (i.e. material, organizational and human resources required

Table 3 Building as a SoPSS requirements definition framework

Service			Product			
Business	Organization	Resources				
Business	Service	People	Equipment	IT System	Energy	Building
High level of abstraction	→					Low Level of Abstraction

Table 4 How disciplines deals with artefacts composing the defined building system

Disciplines	Service			Product		
	Business	Organization	Resource			
	Business	Service	People	Equipment	IT System	Building
Mech. Eng	Ø	+	Ø	+++	Ø	Ø
Syst. Eng	Ø	+	Ø	+++	++	Ø
Soft. Eng	+	++	++	+	+++	Ø
Ent. Arch	+++	+++	++	+	+	Ø
Ind. Eng	Ø	+++	++	+++	Ø	+
Archit	Ø	+	++	++	Ø	+++
Proposal	++	+++	++	+	+	+++

to deliver the service) in a more complex system. Based on similarities between their respective disciplines, this transdisciplinary research points toward finding more holistic outcomes (e.g. a transdisciplinary ontology for requirements definition) to deal with all of these sub-systems in the same framework. Thereafter, relations and dependencies are strengthened toward synergies by adding meaning and information about the complex system.

5 Conclusion

In this paper, the scope of a building is enlarged to the requirements definition of an integrated system. This new system suits well with *transdisciplinary* research. No single domain, especially Architecture, is able to fully define it. Most of them provide tools and techniques to represent and describe each component but independently. This *transdisciplinary* research is based on the concept of Product-Service System that introduces components integration and alignment. Due to their very different kinds, an alliance of discipline is suggested.

At present, "building systems" are defined with less integration between their physical part and service part than potentially expected. The main input comes from the building and ratios instead of the service to provide. Sakao underlines in Sakao and Lindahl (2012) that companies develop services after product realization or even after release on the market. This is the same in construction. Service, equipment, people are refined on the later stages depending on the defined building. As a result, issues appeared after implementation and required changes or brought dysfunctions in the service provision. The proposed mindset called for deeper introspection by the customer on its service and "building system". It would help to anticipate issues with the architects' proposals and it would lead to more aligned proposals with services to provide.

This changed way of seeing a building provided new perspectives of innovation in its definition and a better alignment between all of its components for the design phase. It called for new methods for requirements management (Mauger and Kubicki 2013) in the requirements definition phase but also new methods and practices for the design of buildings. The main effort in the Built Environment concerns computer-aided design and construction phases. The Building Information Modeling (BIM) brings first elements of solution on a multi-disciplinary level through the Industry Foundation Classes (IFC), a neutral and open data model for Architecture-Engineering-Construction (2004). This data model ensures the consistency of the different viewpoints on the same system to design and build in a collaborative environment. However, a few semantic issues remain to be tackled (Eastman et al. 2010). Our research work aims to propose a requirements definition framework that would extend further the BIM development toward earliest phase of conceptual design. *Transdisciplinary* research supports this view. More time and prospect would be required for the requirements definition of a building. The proper alignment would be ensured by a deepened definition of the business and services to be provided.

The conceptual definition of "building systems" is an ongoing project, first sketches of enriched taxonomy is proposed in Mauger (2012). A comparison with existing taxonomies from the other disciplines could lead to the development of trans-disciplinary ontology to be developed later. Thereafter, models from the different disciplines could be applied or adapted (Mauger and Kubicki 2013). The result of such a research would be a systematic and systemic approach for the Architecture discipline.

Acknowledgements The present project is supported by the National Research Fund, Luxembourg.

References

Baines, T. S., Lightfoot, H. W., Evans, S., Neely, A., Greenough, R., Peppard, J., Roy, R., Shehab, E., Braganza, A., Tiwari, A., Alcock, J. R., Angus, J. P., Basti, M., Cousens, A., Irving, P., Johnson, M., Kingston, J., Lockett, H., Martinez, V., Michele, P., Tranfield, D., Walton, I. M., Wilson, H.: State-of-the-art in Product-Service Systems. Proceedings of the Institution of Mechanical Engineers Part B Journal of Engineering Manufacture. pp. 1543–1552 (2007)

Bertelsen, S.: Construction as a complex system. In: Proc. of the 11th IGLC. p. 13, Blacksburg, VA, USA (2003)

Chynoweth, P.: The Built Environment Interdiscipline: A Theoretical Model for Decision Makers in Research and Teaching. In: CIB W089 BEAR 2006. p. 12 (2006)

Cook, M.B., Bhamra, T. a., Lemon, M.: The transfer and application of Product Service Systems: from academia to UK manufacturing firms. Journal of Cleaner Production. 14, 1455–1465 (2006)

Demirkan, H., Kauffman, R. J., Vayghan, J. A., Fill, H. -G., Karagiannis, D., Maglio, P. P.: Service-oriented technology and management: Perspectives on research and practice for the coming decade. Electronic Commerce Research and Applications. 7, 356–376 (2008)

Eastman, C., Teicholz, P., Sacks, R., Liston, K.: BIM Handbook: A Guide to Building Information Modeling for Owners, Managers, Designers, Engineers and Contractors. John Wiley & Sons Ltd (2011)

Eastman, C. M., Jeong, Y., Sacks, R., Kaner, I.: Exchange model and exchange object concepts for implementation of national BIM standards. Journal of Computing in Civil Engineering. 24, 25–35 (2010)
Goedkoop, M. J., Van Halen, C. J., Riele, H. R., Rommens, P. J.: Product Service systems, Ecological and Economic Basics. (1999)
Hussain, R., Lockett, H., Annamalai Vasantha, G. V.: A framework to inform PSS Conceptual Design by using system-in-use data. Computers in Industry. 63, 319–327 (2012)
ISO: Product data representation and exchange ISO 10303–11, (2004)
Jantsch, E.: Towards Interdisciplinarity and Transdisciplinarity in Education and Innovation. In: Apostel, L., Berger, G., Briggs, A., and Michaud, G. (eds.) Interdisciplinarity: Problems of Teaching and Researching in Universities, OECD. pp. 97–121 (1972)
Kalay, Y. E.: Enhancing multi-disciplinary collaboration through semantically rich representation. Automation in Construction. 10, 741–755 (2001).
Kossiakoff, A., Sweet, W.N., Seymour, S.J., Biemer, S.M.: Systems Engineering—Principles and Practice. Wiley-Interscience (2011)
Kubicki, S., Bignon, J.-C., Halin, G.: Assistance to building construction coordination–towards a multi-view cooperative platform. Information Technology in Construction. pp. 565–586 (2006)
Manzini, E., Vezzoli, C.: A strategic design approach to develop sustainable product service systems: examples taken from the "environmentally friendly innovation" Italian prize. Journal of Cleaner Production. 11, 851–857 (2003)
Max-Neef, M. A.: Foundations of Transdisciplinarity. Ecological Economics. 53, 5–16 (2005)
Maussang, N.: Méthodologie de conception des systèmes produits-services, (2008)
Mauger, C., Kubicki, S.: A Conceptual Model for Building Requirements Processing. Proc. of the 11th IPGRC. pp. 318–330. , University of Salford, UK (2013)
Mauger, C.: Method for the conceptual phase of an Integrated Product and Service Design applied to Construction Project. In: Dörr, J., Fricker, S., and Paech, B. (eds.) Proc. of the 18th REFSQ, Vol. 2. pp. 365–372. , Essen, Germany (2012).
Meier, H., Roy, R., Seliger, G.: Industrial Product-Service Systems — IPS 2. CIRP Annals—Manufacturing Technology. 59, 607–627 (2010)
Mont, O.: Product-Service Systems. Final Report. (2000)
Moritz, S.: Service Design—Practical access to an evolving field. London (2005)
Nicolescu, B.: Transdisciplinarity—past, present and future. Congresso Mundial de Transdisciplinaridade. pp. 142–166 (2005)
Ramadier, T.: Transdisciplinarity and its challenges: the case of urban studies. Futures. 36, 423–439 (2004)
Sakao, T., Lindahl, M.: A value based evaluation method for Product/Service System using design information. CIRP Annals—Manufacturing Technology. 61, 51–54 (2012)
Shimomura, Y., Hara, T., Arai, T.: A unified representation scheme for effective PSS development. CIRP Annals—Manufacturing Technology. 58, 379–382 (2009)
Spohrer, J., Maglio, P. P., Bailey, J., Gruhl, D.: Steps Toward a Science of Service Systems. Computer. 40, 71–77 (2007)
Turner, M., Budgen, D., Brereton, P.: Turning software into a service. Computer. 36, 38–44 (2003)
Wiesmann, U., Biber-Klemm, S., Grossenbacher-Mansuy, W., Hadorn, G. H., Hoffmann-Riem, H., Joye, D., Pohl, C., Zemp, E.: Enhancing transdisciplinary research: a synthesis in fifteen propositions. In: Gertrude Hirsch Hadorn, Hoffmann-Riem, H., Biber-Klemm, S., Grossenbacher-Mansuy, W., Joye, D., Pohl, C., Wiesmann, U., and Zemp, E. (eds.) Handbook of Transdisciplinary Research. pp. 433–441. Springer (2008)

5 Conclusion and Outlook

In view of demographic ageing the importance of products tailored to the actual needs and capabilities of people suffering from physical impairments will grow without doubt. The paradigm of human-centered design can help to achieve this requirement. However it can only be applied successfully in collaboration with human scientists since designers usually lack the specialized knowledge necessary to keep track of the inseparable system formed by the product and its user.

In this contribution biomechanical modeling was proposed as a unifying element of transdisciplinary research activities within human-centred design. The development of biomechanical human models that consider human performance restrictions requires a strong collaboration of disciplines like mechanical engineering, anatomy, control system theory, neurology and gerontology. To designers these models offer the unique possibility to analyze and optimize user-product interaction processes without the need to conduct time consuming experiments involving physical prototypes and test persons. But also the more principle oriented human-sciences benefit from having a possibility to apply their findings to a practical problem which can be understood as a validation of hypotheses. Moreover the model may also be seen as a common mean of communication that that enables researchers of different disciplines to share their ideas. The latter is an important element of transdisciplinarity.

The biomechanical simulation approach presented in this contribution is neither complete nor suitable for an application in industry yet. Future work will focus on the implementation of human performance restrictions and the design of a validation process. Nevertheless this example reveals an idea of how product design—especially for people that suffer from physical impairments—can look like in the future.

References

Delp, S., Arnold, C.: OpenSim: Open-Source Software to Create and Analyze Dynamic Simulations of Movement. IEEE Transactions on Biomedical Engineering 54, 1940–1950 (2007)

Feldman, A. G.: Once more on the equilibrium point hypothesis for motor control. Journal of Motor Behavior 18, 17–54 (1986)

ISO 9241–210.: Ergonomics of human-system interaction—Part 210: Human-centered design for interactive systems. International Organization for Standardization, Geneva (2010)

Krüger, D., Stockinger, A., Wartzack, S.: A haptic based Hybrid Mock-Up for mechanical products supporting human-centered design. In: International Conference on Engineering Design, Kopenhagen (2011)

Kimura, F., Yamane, N.: Haptic environment for designing human interface of virtual mechanical products. CIRP Annals—Manufacturing Technology, 127–130 (2006)

Latash, M.: Evolution of motor control: From reflexes and motor programs to the equilibrium-point hypothesis. Journal of Human Kinetics 19, 3–24 (2008)

Maguire, M.: Methods to support human-centered design. International Journal of Human-Computer Studies 55, 587−634 (2001)

Polit, A., Bizzi, E.: Characteristics of motor programs underlying arm movements in monkeys. Journal of Neurophysiology 42, 183–194 (1979)

Rasmussen, J., Dahlquist, J., Damsgaard, M., de Zee, M., Christensen, S. T.: Musculoskeletal modeling as an ergonomic design method. In: International Ergonomics Association XVth Triennial Conference, Seoul (2003)
Reinicke, T.: Potentials and limitations of user integration in product development. PhD thesis, Technische Universität Berlin (2004)
Seeger, H.: Design technischer Produkte, Produktprogramme und -systeme: Industrial Design Engineering. Springer, Berlin (2005)
Stöber, C., Williger, B., Meerkamm, H., Lang, F. R.: Leitfaden für die alternsgerechte Produktentwicklung. Fraunhofer Verlag, Stuttgart (2012)
VDI 2242, Part 2: Engineering design of products in accordance with ergonomics: work aids and relevant literature. Beuth Verlag, Berlin (1986)
Wickson, F., Carew, A., Russell, A.: Transdisciplinary research: Characteristics, quandaries and quality. Futures 38, 1046–1059 (2006)

Management and Collaboration in Transdisciplinary Design

Transdisciplinary Design: The Environment for Bridging Research Across and Beyond Design as a Discipline

Sonia da Silva Vieira

Abstract The present paper elaborates on the notions of discipline and transdisciplinary design to unfold perspectives in design research. A review of the literature and of hints from the observation and empirical studies based on interdisciplinary and transdisciplinary design environments have brought insights into the influence of design. The present contribution proposes an inductive Three Spaces Model of the influence of design across and beyond its discipline. Design has characteristics that constitute different dynamic spaces of influence, namely: transdisciplinary, partly shared and particular. Design host disciplines overlap in shared spaces and particular influence is visible in kernel approaches and resulting artefacts. Invariants of design specific or non-design specific nature have a transdisciplinary influence. Designers, researchers and educators ought to identify such characteristics in order to be able to manage these actions and cope gainfully with the social process of design research, education and practice.

Keywords Design · Particular · Shared · Transdisciplinary · Influence · Model

1 Introduction

No matter what the field of study, a discipline entails a certain activity of experience that provides mental and physical training and its advance into a branch of knowledge (i.e. design). Attempts to define design as a discipline (Archer 1979) delineate from the distinct concerns, values and purposes of design as a domain of extensive influence, and from the need to organize and structure design knowledge in a scientific way (Cross 2001; Love 2002).

Relevant attempts have called for a Design Science (Cross 2001; Simon 1969; Fuller and McHale 1963), however the difficulties in assessing the core knowledge

S. da Silva Vieira (✉)
Institute of Science and Innovation in Mechanical and Industrial Engineering, Faculty of Engineering, University of Porto, Porto, Portugal
e-mail: sonia.dasilvavieira@polimi.it

Politecnico di Milano, Milano, Italy

L. Blessing et al. (eds.), *The Future of Transdisciplinary Design*,
https://doi.org/10.1007/978-3-319-06382-9_13

of design and to achieve an organized and structured body of knowledge have been properly stated (Love 2002; Visser 2009). The maladjustment between the advance of the design practice and design research has been recognized (Gericke and Blessing 2012). In addition, the academic design methodology shows particular incongruity when applied to current or even future design work (Birkhofer 2011). The present re-emergence of Design Science concerns (Cross 2001) is once again asking for the understanding of commonalities and differences of designing across its host disciplines (Love 2002), designers and situations (Fuller and McHale 1963), and beyond the design discipline itself. The call for extending the design practice to other areas of Science with interest in design underlying characteristics (Boland and Collopy 2004) revealed additional difficulties in the transference of the design knowledge beyond its discipline. The transference of design defined as a thought process that can be acquired and embedded through personal development and experience, extensive to all the fields of human action, is still a difficult task for design researchers, educators and practitioners.

Transdisciplinary design emerged as the beacon to understand the practice and the body of knowledge of design across and beyond its discipline. The current perspectives on transdisciplinary design implications are threefold:

1. The transference of non-consensual knowledge in design research;
2. Such transference implies the translation of design essentials, valuable to each area of application;
3. Development of new competencies, roles, research and experimental methods.

As much of this latent knowledge is underdeveloped, some researchers and practitioners are concerned about a certain loss of cohesion of the design discipline and of an excessive adoption of additional tasks to design.

The prefix *trans-* means *across*, *beyond*, *on* or *to the other side of*, *through*, *into another state or place*, *transform*, *translate*, *surpassing*, *transcending*. The adjective *disciplinary* concerns and reinforces the concept of discipline, defined as an activity of experience that provides mental and physical training and its advancement into a branch of knowledge (definitions from the New Oxford American Dictionary). In this paper, design as a discipline is unfolded into three interconnecting spheres of activity and knowledge, namely:

- Design as a Scientific discipline and as a branch of scientific knowledge that attempts to structure, organize, systematize and improve the knowledge and understanding about design through generalized and suitable methods of research, specific design research methods, and methods from other fields, such as neuroscience.
- Design as an Academic discipline, which for purposes of education, is developed in several branches of knowledge, such as Design Culture (Julier 2008) and Design Experience (Schifferstein and Hekkert 2008).

- Design as a Practical discipline, rooted in several professional and core practices of design, such as the host disciplines (Love 2002) of Graphic, Interaction, Industrial, Architecture, Engineering, Software and Services design, that subsequently unfold into many specializations and emergent interests.

As *'Design constitutes human being'* (Krippendorff 2006) this paper assumes that transdisciplinary design constitutes:

4. The design characteristics that are broadly shared and are, therefore, invariant across design host disciplines, designers, situations, and beyond the design discipline.
5. The design environments that integrate multidisciplinary activities (i.e. design, biology, literature, marketing, management, neuromarketing, creativity) and interdisciplinary ones (i.e. industrial design, architecture, mechanical engineering, as host design disciplines).
6. An attempt to bring new perspectives and developments to assess, organise and structure the results obtained from design research.

Transdisciplinary design brings a reinforced call for the intervention of design, for example, through action research in unfamiliar areas, or creating the context for the convergence of knowledge from other fields (i.e. neuroscience and creativity) to ensure high reliability of the research. Therefore, transdisciplinary design environments are appropriate to explore and bridge research across and beyond design. This paper, proposes a Three Spaces Model of the influence of design across and beyond its discipline.

2 Review of the Literature

A Generic-design hypothesis was proposed in the past (Goel and Pirolli 1989, 1992) regarding the study of design as a subject matter in its own right. The study presents findings of commonalities in the structure of design problems and tasks across some design disciplines, and contributes to structuring the core knowledge of design. This attempt was augmented to a cognitively oriented generic-design hypothesis (Visser 2009) that structures avenues for research. Several researchers attempts have demonstrated similarities and differences across design and non-design disciplines. Relevant aspects have been addressed, such as similarities and differences of cognitive processes in different fields of design (Akin 2001; Gero 2010), differences in terminology across design disciplines (Reymen 2001), designers particular forms of knowledge (Cross 2006), the artefacts characterizing the design discipline as the variant underlying the differences across disciplines (Hubka and Eder 1987), among many other, though somewhat dispersed contributions. Table 1 illustrates results from the literature.

Table 1 Overview of Variants, Invariants and design specific characteristics

Design Invariants
1. Design is a type of cognitive activity rather than just a professional status (Simon 1969; Jones 1979)
2. Design is a problem-solving activity (Newell and Simon 1972) non-consensual
3. Design problems are considered ill-defined (Simon 1973; Thomas and Carroll 1979)
4. Problem space and solution space progress in co-evolution (Dorst and Cross 2001)
5. Design is a 'satisficing' activity (Simon 1987)
6. Design involves complex problems that are rarely decomposable into independent sub-problems (Akin 2001; Simon 1973; Goel 1995) non-consensual
7. Designers generate an initial solution kernel (Cross 2001; Darke 1979)
8. Design problems have several acceptable solutions (Eastman 1970)
9. Solutions, evaluation criteria and procedures undergo evaluation (Lera 1981; D'Astous et al. 2004)
10. Designers re-use of knowledge through analogical reasoning (Casakin and Goldschmidt 1999)
11. Design activity is mostly opportunistically organized (Visser 2009,1994)
12. Invariants of design cognition (Akin 2001; Gero 2010)
Design Variants
1. Process and organization
2. Tools in use
3. The role of the user in the design process
4. Designer
5. Expertise
6. Routine tasks
7. Idiosyncrasy
8. Social embeddedness (Visser 2009)
9. Variants in design cognition (Akin 2001)
10. Artefacts' evolution and types of artefact (Hubka and Eder 1987)
Design Specific
1. Specific cognitive activities and structures (Visser 2009)
2. Designers have specific forms of knowledge (Cross et al. 2002)
3. Expertise in design is specific. A key feature is 'problem framing' (Cross 2004)
4. Expert designers are solution-focused rather than problem-focused (Lawson and Dorst 2009)
5. There exists a designerly way of thinking and communicating (Archer 1979; Cross 2006)

However, the later generic-design hypothesis (Visser 2009) makes claims for a wider and augmented empirical evidence, validation and understanding of the relationships between these characteristics. Such understanding requires consensual and generalized clarity around the borders of the host disciplines, terminology, concepts, contexts and situations of designing (Love 2002). Figure 1 sums up the current call for empirical evidence of design characteristics.

The structure of the call for empirical evidence concerns the following assumptions:

- Different design host disciplines share major commonalities.
- Different design host disciplines show variation in similarities.
- Designers' approaches vary across and within design host disciplines.

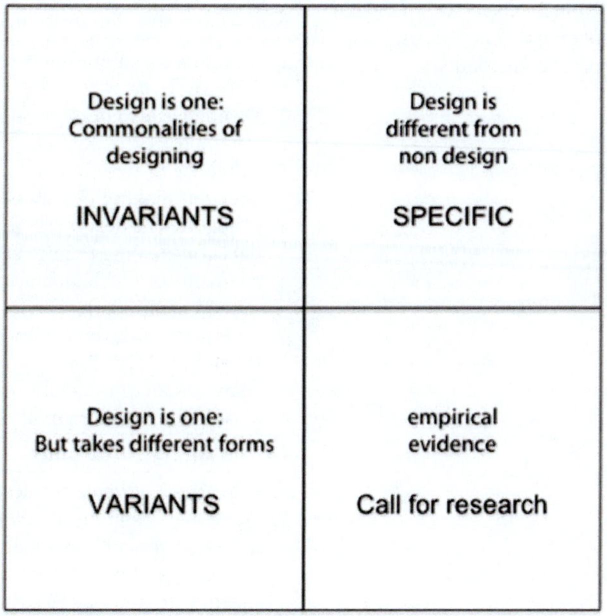

Fig. 1 Overview of the call for empirical evidence of variant, invariant and specific characteristics of design. Based on Goel and Pirolli 1992, Akin 2001, and Visser 2009

- Design situations create variance in designers' cognition.
- Design is different from non-design.

Real life design practice has been recognized as the appropriate environment to empirically assess designing. Shared methods of research have been proposed to improve the assessment of designing characteristics to a significant level for theory building (Gero 2010). However, neither consensus nor platforms have been established for the assessment, exchange, validation and replication of these studies.

This paper argues that a different perspective might help in restructuring the call for empirical evidence. Such a perspective is based on the literature review and on hints from empirical studies further described:

1. Some invariants of design are not design specific.
2. What is specific about design is also invariant across and beyond its discipline.

3 Hints from Multidisciplinary and Transdisciplinary Design Environments

The analysis of results from previous research (Table 2) based on empirical studies brought into evidence some relevant features of design characteristics. Observation periods in four design consultancies provided the social context and the opportunity to observe and assess designers' characteristics. The design consultancies were based on the host disciplines of graphic and interaction design, architecture and mechanical engineering. Some of these design environments were interdisciplinary, such as the

Table 2 Overview of design specific and non-design specific invariants

Non-design specific invariants
1. Design issues iteration process characteristics (Vieira et al. 2011a) 2. Design issues interdependency process characteristics (Vieira et al. 2011a) 3. Influence of iteration and interdependency in decision-making (Vieira et al. 2011a) 4. Categories of designers' drivers of priority value in decision-making (Vieira et al. 2010a) 5. Elements of mechanisms to cope with crisis situations (Vieira et al. 2010b) 6. Invariants of designers and non-designers approaches (Vieira et al. 2010c) 7. Flow model of interaction in design meetings with stakeholders (Vieira et al. 2011b)
Design specific invariants
1. Categories of value for designers' (Vieira et al. 2009) 2. Advantages and disadvantages of using structured methods as design procedures (Vieira et al. 2011c) 3. Categories of priority issues for decision-making in design meetings (Vieira 2013) 4. A Meta-level behaviour framework for critical situations and crucial actions in design (Vieira et al. 2012) 5. Categories of characteristics of designers' approaches (Vieira et al. 2010c)

architecture and graphic design case studies, with external multidisciplinary interactions, while the interaction design and mechanical engineering case studies had a transdisciplinary design environment.

Data was assessed through observation of on-going projects and interviews of people with different backgrounds and activities, from graphic, industrial, interaction, editorial and communication design, to social communication, literature, illustration, biology, architecture, mechanical, civil, environmental, electronics and aerospace engineering, to project management and sales.

Studies were focused on the identification of variants and invariants of design for specific purposes and research objectives. Invariants of design are difficult to assess and their recognition is derived from the analysis and descriptions of successful human behaviour (Simon 1990). The progress in research and the understanding of the different case studies, cultures, design philosophies, specific methods, tools and sequence of practices has led to the following four aspects:

1. Design across disciplines has different levels of influence;
2. Generally assumed design characteristics are widely shared among designers and non-designers;
3. Design beyond its discipline has a level of influence based on invariant and design-specific characteristics;
4. Invariants are achievable at an abstract and qualitative level.

Designers aim for the implementation of design. Such purpose drives designers to design action in concrete situations. Such activities and situations happen in the social context of design within its multiple, interdependent and multidirectional axis. Value judgment analysis and decision-making are invariants of designers' behaviour that play a crucial and interdependent role in design as a social and individual process. However, such invariants are not design specific. These and other characteristics are summarized in Table 2.

Research shows that not all the invariants are design specific. However, design specific characteristics are also invariant, but more difficult to identify as unique design characteristics. Results in variant and particular characteristics of host disciplines with influence on the design approach were also identified (Vieira et al. 2010c). Whether invariants are design specific or not, they have a transdisciplinary influence, especially those that constitute behaviour in exchange and negotiation processes, of interdependency and decision-making. Design specific invariants relate to processes of evaluation, procedural flexibility, design issues and characteristics of design approach. What is transdisciplinary, is not just invariant across or beyond the design discipline, but also asks for clear identification, transference and translation, focused on essential aspects for each area of application. Variants of design can have a partially shared or particular influence across designers, design environments and host disciplines. Some variants have particular influence and characterize the design host discipline or the design approach. Other variants, partially shared among designers, relate to background activity and affinity. Figure 2 illustrates the transdisciplinary, partially shared and particular spaces of the influence of variants,

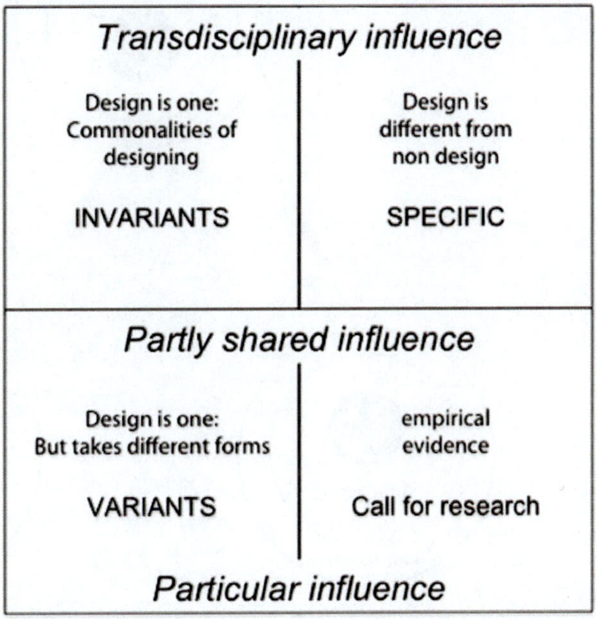

Fig. 2 Restructuring of the call for empirical evidence in design research

invariants and specific characteristics of design, and proposes the restructuring of the call for empirical evidence in design research.

The case study based on a small medium enterprise of interactive design solutions revealed additional characteristics which are absent in small design consultancies, but present in large companies. The mutual influence and exchange processes between sales, marketing, research, design and product development departments has provided the appropriate environment to understand the need to extend the borders of design research to a transdisciplinary research approach. Understanding design-underlying processes in transdisciplinary design contexts can help in identifying design invariant and specific characteristics. Such identification can better support the extension of the practice of design in other areas of human activity thus overcoming the present difficulties of design knowledge transference beyond its discipline.

4 Discussion

An inductive Three Spaces Model of the influence of designing across and beyond its discipline (Fig. 3) is derived from the review of literature and hints from empirical studies. Design as a discipline has three spheres of influence, namely, scientific,

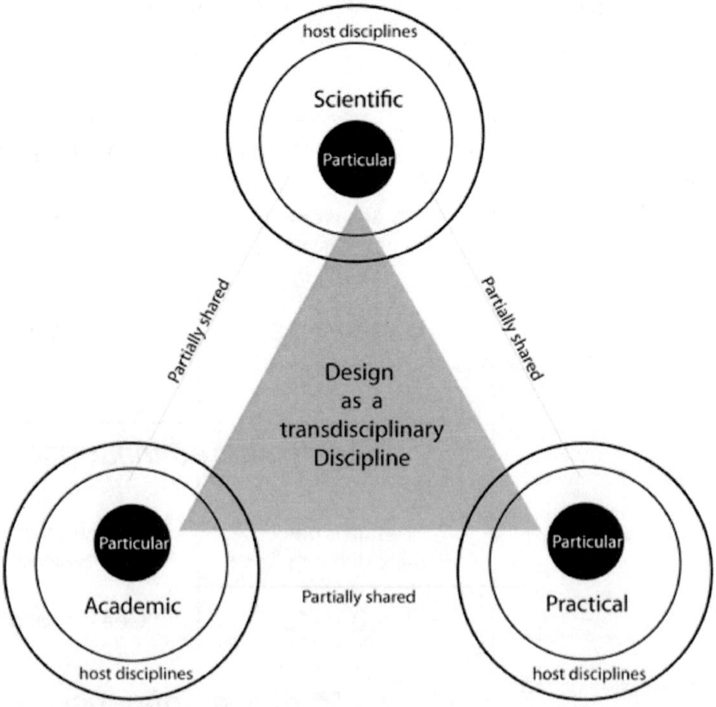

Fig. 3 Three Spaces Model of the Influence of Design

academic and practical. Each one of these spheres hosts different disciplines. The interaction between the agents of each host discipline asks for the identification and exchange of design essentials, therefore specific and invariant characteristics. This dynamic platform is the transdisciplinary space of the influence of design. Each sphere has influence over particular characteristics of design that relate to the practical, academic and scientific design contexts, as well as to particulars of each host discipline. Partially shared spaces also have higher dynamics between spheres of influence and host disciplines.

Characteristics of each of the three spaces of the influence of design are described, namely:

1. Transdisciplinary space. For example, the extension and consequent absorption of new technology, rules and policies that asks for new design approaches in classic design disciplines. Translation of design specific characteristics into another state for exchange, expansion and inclusiveness of transdisciplinary design. Change of methods, thinking patterns and the ability to communicate across design and non-design professions with different perspectives of the design processes.
2. Partially shared space. For example, characteristics shared by groups of designers, such as engineering-based designers (Mechanical, Software, industrial design), Art-based designers (Graphic, Architects), Design-based designers (interaction, experience, product design) that frame and influence design approaches and results in interdisciplinary and multidisciplinary environments.
3. Particular space. For example, host discipline characteristics, culture, particular design problems, design situations and characteristics of design approaches. How far the designing of the intangibles (Jones 1979) entails particular characteristics per design discipline. The creation of emergent and still unstructured design disciplines, such as interaction, control and service design of interdisciplinary influence.

Design, as an activity, has characteristics that result in different levels of influence across disciplines, namely: transdisciplinary, partially shared, and particular. Design disciplines overlap each other in common and partially shared spaces, and particular influence is visible in kernel approaches and resulting artefacts that characterize different design disciplines. In essence, design entails mental and physical actions that can be transdisciplinary, partially shared or particular to designers across and beyond its discipline. Designers, researchers and educators ought to identify such characteristics in order to manage these actions and cope gainfully with the social process of design research, education and practice.

5 Conclusion

The present study concludes with avenues to explore and research how far the knowledge that derives from design studies can be extended to other disciplines, and fields of sciences in a transdisciplinary dimension. From the studies of design as a whole and the units of design, comparison occurs in a wider, interconnected, social, political and business context of mutual and influencing forces. Such context requires a platform with a more conscious understanding supported by leading developments in design methodology, enabling a design guidance that is flexibly applicable across areas where design can have an influence.

This contribution proposes the creation of a research platform in the transdisciplinary space, constituted by interconnected research groups based on different design host disciplines. Research in transdisciplinary issues per host discipline, can further the identification of inter, multi and transdisciplinary problems, difficulties, barriers, research gaps, undeveloped roles and methods. Periodical events would allow sharing such concerns and delineate strategies for research in transdisciplinary issues across the design discipline. On the implementation level, such a platform, which is aimed at design practice and research, would benefit from the adoption of action research as an approach that could have an influence in the contexts of intervention, and also benefit from the creation of seeds for knowledge development.

Acknowledgements The author gratefully acknowledges Atelier Henrique Cayatte, João Álvaro Rocha Architects, Ydreams, the Delft Center for Systems and Control for empirically supported previous research.

References

Akin, Ö.: Variants in design cognition. In: C. Eastman, M. McCraken and W. Newstetter (eds.) Design Knowing and Learning: cognition in design education, Elsevier, Amsterdam (2001)

Archer, L.: Design as a Discipline. Design Studies, 1(1), 17–20 (1979)

Birkhofer, H.: Introduction, In: Herbert Birkhofer (Ed.), The Future of Design Methodology. Springer, London (2011)

Boland, R., Collopy, F. Design Matters for Management. in Boland, R. and Collopy, F. (eds.) Managing as Designing, Stanford, California, Stanford University Press (2004)

Casakin, H, and Goldschmidt, G.: Expertise and the use of visual analogy: Implications for design education. Design Studies, 20:153–175 (1999)

Cross, N.: Designerly ways of knowing: design discipline versus design science. Design Issues, 17(3), 49–55 (2001)

Cross, N.: Designerly ways of knowing. London, Springer-Verlag (2006)

Cross, N.: Creative cognition in design: processes of exceptional designers. In: Hewett, Tom and Kavanagh, Terence (eds) Creativity and cognition. New York, USA: ACM Press, pp. 14–19 (2002)

Cross, N.: Expertise in design: an overview. Design Studies, 25(5), pp. 427–441 (2004)

Darke, J.: The primary generator and the design process. Design Studies 1(1), 36–44. (1979)

D'Astous, P., Détienne, F., Visser, W., and Robillard, P.N.: Changing our view on design evaluation meetings methodology: a study of software technical review meetings. Design Studies, 25, 625–655. (2004)

Dorst, C., Cross, N.: Creativity in the design process: co-evolution of problem-solution. Design Studies 22, 425–437 (2001)

Eastman C. M.: On the Analysis of Intuitive Design Processes, Emerging Methods in Environmental Design and Planning. G. T. Moore (ed), MIT Press, Cambridge, Ma. (1970)

Fuller, R. B., McHale, J.: World design science decade, 1965-1975: Five two-year phases of a world retooling design proposed to the International Union of Architects for adoption by world architectural schools. Carbondale: World Resources Inventory. Southern Illinois University at Carbondale (1963)
Gericke, K., Blessing, L.: An Analysis of design process models across disciplines. In: Design Conference, Dubrovnik, Croatia (2012)
Gero, J.: Generalizing Design Cognition Research. In: Proceedings of DTRS 8, Sydney, Australia (2010)
Goel, V.: Sketches of Thought. Bradford Bks. MIT Press, Cambridge, Ma. (1995)
Goel, V., Pirolli, P.: Motivating the Notion of Generic Design within Information-Processing Theory. The Design Problem Space, AI Magazine 10 (1), 18–36 (1989)
Goel, V., Pirolli, P.: The Structure of Design Problem Spaces. Cognitive Science, 16, 395-429 (1992)
Hubka, V., Eder, W.: A scientific approach to engineering design. Design Studies, 8(3), 123–137 (1987)
Jones, J. C. Designing Designing. Design Studies 1(1), 31–35 (1979)
Julier, G.: The Culture of Design. Sage (2008)
Krippendorff, K.: The Semantic Turn, A New Foundation for Design. Boca Raton, FA: CRC Press, Taylor & Francis Group, p. 74 (2006)
Lawson, B., Dorst, K.: Design Expertise. Oxford, Auckland, Boston, Johannesburg, Melbourne, New Delhi, Elsevier (2009)
Lera, S.: Architectural designers' values and the evaluation of their designs. Design Studies, 2(3), 131–137 (1981)
Love, T.: Constructing a coherent crossdisciplinary body of theory about designing and designs: some philosophical issues. Design Studies, 23, 345–361 (2002)
Newell, A., Simon, H. A.: Human problem solving. Englewood Cliffs, NJ: Prentice-Hall (1972)
Reymen, I.: Improving Design Processes through Structured Reflection, A domain-independent approach, Technical University of Eindhoven, Eindhoven (2001)
Simon, H. A.: The Sciences of the Artificial. MIT Press, Cambridge, MA. (1969)
Schifferstein, H. N. J., Hekkert P.: Product Experience, Elsevier (2008)
Simon, H. A.: The structure of ill-structured problems. Artificial Intelligence, 4, 181–201 (1973) also in Cross, N. (ed.). Developments in design methodology. Chichester, England: Wiley. 1145–1166 (1984)
Simon, H. A.: Problem forming, problem finding, and problem solving in design. In A. Collen & W. W. Gasparski (eds.), Design and systems: General applications of methodology. vol. 3, 245–257 (1987)
Simon, H. A.: Invariants of Human Behavior. Annual Reviews Psychology, 41, 1–19 (1990)
Thomas J., Carroll, J.: The psychological Study of Design. Design Studies 1(1), 5–11 (1979)
Visser, W.: Organisation of design activities: Opportunistic, with hierarchical episodes. Interacting with Computers, 6(3), 239–274 (1994)
Vieira, S., Badke-Schaub, P., Fernandes, A., Fonseca, T.: Designers' thinking and acting in design meetings. In: International Association of Societies of Design Research Conference, IASDR, 30 October–4 November 2011, Technical University of Delft (2011)
Vieira, S., Badke-Schaub, P., Fernandes, A., Fonseca, T.: Understanding how designers' thinking and acting enhance the value of the design process. In: Design Thinking Research Symposium, DTRS8, Sydney, Australia (2010)
Vieira, S., Badke-Schaub, P., Fernandes, A., Fonseca, T.: Replacing routine: reframing design opportunities for creativity and innovation. In: 16th International Product Development and Management Conference, 13–15 June, Murcia, Spain (2010)
Vieira, S., Badke-Schaub, P., Fernandes, A., Fonseca, T.: Substance variation in design approach. In: Design Conference, Dubrovnik, Croatia (2010)
Vieira, S., Badke-Schaub, P., Fernandes, A., Fonseca, T.: Designers' thinking and acting in meetings with Clients. In: International Conference on Engineering Design, ICED11, 15–18 August 2011, Technical University of Denmark (2011)

Vieira, S., Badke-Schaub, P, Fernandes, A., Fonseca, T., Cardoso, C.: Translating Lean Thinking to design: Identifying Muda in designing. In: 1st International Conference on Integration of Design, Engineering, and Management for innovation14–15 September, Porto (2009)
Vieira, S., Fernandes, A., Bruens, G.: Educational Context Impact on New Product Development Learning Experience. In: 17th International Product Development and Management Conference, IPDMC, 5–7 June, Delft, Netherlands (2011)
Vieira, S.: Crucial actions in design, a framework of awareness to critical situations. From a Lean Thinking perspective. Doctoral Thesis. Technical University of Delft, Delft (2013)
Vieira, S., Fernandes, A., Badke-Schaub, P., Fonseca, T.: Sources of critical situations as crucial actions in design. In: Design Conference, Dubrovnik, Croatia (2012)
Visser, W.: Design: one, but in different forms. Design Studies, 30(3), 187–223 (2009)

A Comparison of Two Transdisciplinary Human-Centered Design Approaches for Poverty Alleviation

Jessica Vechakul and Alice Agogino

Abstract This exploratory study characterizes two transdisciplinary human-centered design approaches for creating novel products or services for poverty alleviation. *Transdisciplinary design* involves the integrated use of tools, techniques, and methods from multiple disciplines in one holistic process. The term *discipline* includes academic fields of study that are taught in universities, as well as specialized expertise that are developed through life experience. Two pioneering organizations were selected to be exemplary case studies based on their high regard and influence within the design industry, social sector, and academia. This paper highlights similarities and differences between the design thinking approach practiced by IDEO.org (a nonprofit design consultancy) and the Creative Capacity Building approach developed by the International Development Design Summit (an educational organization hosting annual innovation conferences). IDEO.org's teams of professionals (e.g., industrial designers or business strategists) develop innovative products and services for implementation by partners serving low-income communities. IDDS teaches people from diverse backgrounds (e.g., farmers, mechanics, students, teachers, doctors, and artisans) to create technologies and launch enterprises for poverty alleviation. The objective is not to determine which approach is better, but to determine what can be learned from IDEO.org about designing with established organizations, and from IDDS about teaching budding innovators to be grassroots change agents.

Keywords Human-centered design · Design thinking · Creative capacity building · Participatory development · Co-creation · Co-design · Appropriate technology

1 Introduction

Socioeconomic development projects are typically designed by experts from a single discipline without involving people who are the intended users of the solution.

J. Vechakul (✉) · A. Agogino
Mechanical Engineering Department, University of California, Berkeley, CA, USA
e-mail: jessvech@berkeley.edu

L. Blessing et al. (eds.), *The Future of Transdisciplinary Design*,
https://doi.org/10.1007/978-3-319-06382-9_14

However, this approach often results in ineffective solutions that neglect social factors, such as cultural preferences or behavior change. With human-centered design (HCD), design decisions are guided by the needs of potential *users* or people whose experiences will be transformed by the design (Gasson 2003; Dym et al. 2005). HCD is especially important for facilitating empathy since development practitioners often come from different cultural and socioeconomic background than their intended users. Moreover, the multidimensional nature of sustainable development requires the expertise of many disciplines, thereby making transdisciplinary design not only an asset but a necessity (Wahl and Baxter 2008; Eisenbart et al. 2012). Rather than members of different disciplines working with separate processes, *transdisciplinary design* involves integrated use of tools, techniques, and methods from multiple disciplines in one holistic process to create a novel product, service, or system meeting a complex societal need (Scholz 2000; Ertas et al. 2003). The relationship between the disciplines is continually evolving based on the needs of the project, with various disciplines blending or leading at different stages. The term *discipline* includes academic fields of study, as well as specialized expertise from life experience. Scholz asserts that knowledge that is intuitive and experiential is just as valid as knowledge that is analytic and abstract (Scholz 2000). For example, farmers and agricultural engineers possess different yet relevant spheres of knowledge. Furthermore, when working across cultural boundaries, knowledge of local language, customs, and social norms should be recognized as expertise.

As HCD is an emerging practice in the social sector, a critical first step is an exploratory study. This paper characterizes two HCD approaches to address challenges in low-income communities. Two pioneering organizations were selected to be exemplary case studies based on their high regard and influence within the design industry, social sector, and academia. This paper compares the design thinking approach practiced by IDEO.org and the Creative Capacity Building (CCB) approach developed by the International Development Design Summit (IDDS).

1.1 Pioneering Organizations in HCD for Social Impact

This exploratory study offers a snapshot of two pioneering organizations when they first entered the field of HCD for Social Impact. Since the completion of this study, both organizations have continued to evolve and refine their approaches and offerings. The scope of this study was limited to the approaches of these organizations during the first few years of their existence; it is not intended to imply or comment upon how these organizations have changed after the completion of the study. What follows is a summary of how each organization was founded and their respective design approaches from 2007 to 2012 for IDDS and from 2011 to 2012 for IDEO.org. IDEO, the award-winning global design firm, is widely renowned for popularizing design thinking. Design thinking is considered to be "potentially universal in scope, because design thinking may be applied to any area of human experience" (Buchanan 1992). In 2011, two former leaders of IDEO's Social Innovation Domain founded IDEO.org as an independent nonprofit organization that works with partner organizations (nonprofits, social enterprises, and foundations) to design for poverty alleviation. For the first few years after IDEO.org launched, IDEO.org recruited a new Fellowship class of IDEO designers and social sector leaders each year. IDEO.org's recruitment criteria for the Fellowship program was somewhat influenced by IDEO's concept of a "T-shaped" profile, represented by a disciplinary depth of skill to make tangible contributions to the team as well as "empathy towards people and disciplines beyond one's own" (Brown and Wyatt 2010). At IDEO.org, transdisciplinarity is represented by a diversity of professional disciplines including industrial design, business strategy, engineering, social marketing, journalism, and information design. In 2011, each of the three IDEO designers led a team of two to three other Fellows working on 6-week to 12-week design projects. Fellows were assigned to teams based on the expertise required to produce the best deliverable for each project. To ensure high-quality design, IDEO creative directors reviewed progress at critical milestones.

The International Development Design Summit (IDDS) was founded in 2007 at MIT, and has been recognized by USAID as a model of excellence for engineering education. For IDDS, transdisciplinarity goes beyond academic disciplines to include expertise from trade skills or life experiences. For the first few years after the first IDDS, there were annual 3- to 5-week conferences that used the Creative Capacity Building (CCB) approach to inspire and enable people to create technologies for poverty alleviation. IDDS design teams included people with a broad range of expertise and experiences (e.g., welders, nurses, religious leaders, and engineers).

With roots in the appropriate technology movement and participatory development, CCB is based on the premise that *anyone* can become an active creator of technology, not just a recipient or user of technology (Taha 2011). The first few conferences organized by IDDS brought together over 60 people from more than 20 countries worldwide to form design teams and innovate livelihood technologies that "increase income, improve health and safety, decrease manual labor or save time" (Taha 2011). These IDDS participants learned the design process through lectures and hands-on workshops, and put these principles and skills into action on team projects. Each team was assigned a mentor, who guides the team based on extensive experience in design or entrepreneurship. IDDS design teams were formed based on the participants' project preferences. Since IDDS highly values diversity in teams, it was common for teammates to speak different languages and have different socioeconomic and disciplinary backgrounds.

1.2 Methods and Study Projects

This exploratory study is a qualitative analysis of documents, in-person observations of design projects, and informal conversations. The lead author worked as an IDEO.org Fellow between September 2011 and May 2012, and served various roles as an organizer, participant, and team mentor for IDDS in 2007, 2008, 2009, 2010, and 2012. Working on design teams on four projects with IDEO.org and five projects with IDDS enabled the lead author to have the perspective of an embedded researcher (Tietje and Scholz 2001). The projects include work in water, sanitation, and alternative energy in Africa and Asia. The deliverables included early functional prototypes, business models, and a brand strategy.

2 The Design Process

Most design processes are linear, iterative, and include the core stages of establishing a need, analysis of a task, conceptual design, embodiment design, detailed design, and implementation (Eisenbart et al. 2012). CCB and design thinking include these core activities, and also adds "gathering information from users" Taha (2011), which is integral to HCD approaches. Both frameworks view the design cycle as a process of iterative refinement with more detail, depth, and understanding gained with each iteration. Although most design processes are linear (Eisenbart et al. 2012), CCB is represented as a cycle (Taha 2011) and design thinking is represented as a system of overlapping spaces (Inspiration, Ideation, and Implementation) (Brown and Wyatt 2010). One can "think of inspiration as the problem or opportunity that motivates the search for solutions; ideation as the process of generating, developing, and testing ideas; and implementation as the path that leads from the project stage into people's lives" (Brown and Wyatt 2010). A design thinking team may be in two or three spaces simultaneously, and the transition through these spaces is not necessarily sequential (Brown 2008).

Contrary to multidisciplinarity, in which each expert in the group may be advocating for his own opinion or process, transdisciplinarity promotes a collective ownership of ideas Brown (2009). Unlike other design approaches in which each discipline serves a unique role, in design thinking and CCB, teams overlap in activities with each individual stretching beyond his expertise to contribute in *all* phases of the design process. For example, instead of an ethnographer interviewing the user and giving insights to a designer who then creates a product, all team members would conduct interviews and create an integrated solution together.

2.1 The Project Brief

A design project typically began with a project brief that established realistic goals with opportunities to explore and discover unexpected and serendipitous solutions. IDEO.org's leadership worked with partners to carefully craft a brief that would maximize the impact of the design team for the partner's goals. Since IDEO.org aims to have measurable impact, projects tended to focus on products or services rather than strategy.

IDDS projects were proposed by IDDS' network of development workers, lecturers, or participants. Although global applicability was a long-term goal, IDDS projects initially focused at the community level with partnerships in city neighborhoods or villages. Projects were selected based on various criteria including the design team's interests, the potential for innovation and scalability, proposals by a founder who had committed to launching a venture, or the needs of partner communities. Project types included fundamental scientific research, adapting an existing technology for low-cost production in developing countries, and modifying crop-processing equipment with alternative power inputs.

Notably, design thinking and CCB both left flexibility for design teams to change the project scope. The possibility of reframing a problem is critical for innovation. For example, an IDDS brief initially focused on designing a device that heats and disinfects breast milk containing HIV. The team redefined the problem as preventing mother-to-child transmission of HIV, and developed a novel concept to chemically deactivate HIV in breast milk as it passes through a nipple shield. The team received funding from the Gates Foundation to research this promising idea. This new problem framing enabled the team to explore an entirely new design space, leading to an innovation that is potentially cheaper, easier to distribute, and more discreet for mothers to use.

2.2 Inspiration

During *Inspiration*, the team gathers information that will improve understanding of the problem and possible solutions. Before starting user research in the field, IDEO.org and IDDS design teams conduct secondary research online about the current situation, competitors, and analogous inspirations from other fields. IDEO.org design teams sometimes received information and guidance from partners or IDEO designers who have worked on similar projects. IDEO.org's and IDEO's vast professional network also gave the design team access to experts from various sectors. Since most of IDEO.org's projects focus at the regional or national scale, user research was conducted in multiple locations to gain a broader understanding of trends beyond a single village or neighborhood. Typically, the design team created a research plan including a rough schedule, methods (e.g., shadowing or semi-structured interviews), user profiles, and interview questions. The team also sometimes created artifacts, photo prompts, or other props that would help a user imagine a scenario. IDEO.org's local contacts or partners helped to organize site visits or user interviews. Sometimes, IDEO.org hired a local market research agency to identify potential users and sometimes compensated users for their participation. Implementation partners were encouraged to accompany the design team during the user research so they could learn directly from users. To gain credibility quickly and facilitate understanding, IDEO.org worked with local partners who serve as translators, cultural guides, and community liaisons (Brown and Wyatt 2010). IDEO's HCD Toolkit included methods (e.g., self-documentation through photos) to help users express what may be tacit knowledge, or that which is implicit or inherently understood but difficult to verbalize (Polanyi 2009).

For IDEO.org, the needs of their implementation partners and users *guided* design decisions, but they were generally not part of the core design team. Some transdisciplinary design approaches assume that full participation by users is ideal (Caruso and Frankel 2010; Arnstein 1969). However, this paper suggests that the organization and project goals dictate whether it is appropriate to involve users and the extent of their participation. For a project designing a brand for a dignified sanitation service, an IDEO.org design team interviewed employees and users to understand what emotional meanings were associated with the service. Since marketing and graphic design are beyond the skill set of an average person, teaching these skills to enable full participation would have required additional resources. In this project, users were appropriately consulted, and the users' values of reliability, comfort, and pride became core service principles.

For IDDS, user research involved a general needs assessment of a village or urban neighborhood to understand the context of users' lives beyond the project. IDDS teams lived in the communities were the projects were based, for periods of a few days to weeks and often participate in many of the users' daily activities. Rapport and trust were developed during this time that would have been difficult to reproduce during a brief interview. IDDS teams also gathered information through observations and interviews.

IDDS specifically encouraged community members (people from the villages or urban areas where project fieldwork was conducted) to participate in IDDS and serve as cultural guides and liaisons to the community. Volunteer translators enabled community members on the design team to fully engage in the design process. It is important to note that although community members could contribute relevant knowledge about the context of use and the intended user, they were sometimes not the intended users. IDDS recognized that no individual can represent the needs of everyone in his community. Although community members' contextual knowledge and practical expertise were valued and respected, IDDS taught participants to question their assumptions and to gather feedback from the intended users.

2.3 *Ideation*

During *Ideation*, design teams generate many concepts, prototype to learn, and select the most promising concepts to implement. Especially for transdisciplinary design teams in which disciplinary terminology may differ widely, *intermediary objects* (representations that are created or manipulated to support integration of knowledge) are crucial for developing a common understanding of the design problem and proposed solutions (Boujut and Blanco 2003). Since IDEO.org designs strategies or services in addition to products, the intermediary object may range from abstract constructs to tangible prototypes. The intermediary object for IDDS is typically a tangible prototype. IDEO.org had dedicated project spaces where the design team could access intermediary objects and interact real-time with them. Synthesis is the process by which design thinkers distill what they observe into insights that can lead to opportunities for change or solutions. Design teams often capture important themes from user research with sticky notes because the limited size encourages conciseness, and the colors and mobility enable rapid categorization and pattern recognition. Brainstorming is a popular method for generating many ideas, which are evaluated through design reviews with feedback from IDEO.org's leadership, clients, or users. Often, rather than choosing one idea over another, the promising elements of various ideas were combined.

Since industrial design was one of the core disciplines at IDEO.org, visual thinking was naturally encouraged. Drawing forces decisions and captures emotional content as well as functional characteristics (Brown 2009). Sketching has also been shown to enable insights and the co-evolution of the design problem and possible solutions. User experiences are prototyped with visualizations or narratives (e.g., personas or storyboards). Physical prototypes can also be fabricated with increasing refinement from sketch modeling materials (foam core, hot glue, etc.) to 3D printed or machined parts.

IDDS teams did not work in a dedicated project space, but they shared workshops with other teams, which facilitated cross-pollination of ideas and collaboration across projects. IDDS teams discussed insights and ideas verbally, but written communication might have been more difficult for teams managing multiple languages and varying levels of literacy. IDDS teams learned to convert user needs into design requirements (e.g., speed, power, cost, etc.) that could be measured and tested with simple experiments. Teams brainstormed ideas and evaluated their concepts against those metrics with Pugh Charts. Learning to use basic hand tools and building a simple functional device (e.g., water pump, solar lantern, etc.) were core components of the CCB curriculum. Found or recycled materials or inexpensive parts like PVC pipes and steel stock were commonly used for prototyping. Building physical prototypes with simple tools and materials facilitated communication and shared understanding across disciplines and cultures.

2.4 *Implementation*

During *Implementation*, ideas move towards realization. As consultants, IDEO.org's impact upon end users was dependent upon whether partners decided to implement the concepts. Consequently, conveying a plausible story of a compelling need and solution to the client is critical. Sometimes, the story itself was the deliverable and the tangible product was sometimes a "slide deck" presentation, which contained insights evoking empathy for users and inspiring ideas. Stories, user profiles, quotes, and pictures were commonly used to convey research findings within an "Insights and Opportunities" framework. Prototypes could be conceptual ideas or looks-like renderings that were meant to capture the imagination. Sometimes, detailed artifacts (e.g., a financial model, customer journey, sample advertisement) served as examples of how a concept could come to life. However, regardless of how promising a concept was or how well its value and actionability were communicated, partners could decide not to implement. The project may no longer be a priority to the organization due to a shift in strategy, change in leadership, or budget constraints. Despite these challenges, 50% of the projects from IDEO.org's first year were implemented.

For IDDS, implementation refers to refinement of a physical prototype, fabrication, testing and evaluation, and gathering user feedback. IDDS conferences ended with a final presentation at a public event, at which community stakeholders are invited to give feedback on the teams' prototypes. Since IDDS focused upon engineering design and innovating early-stage technologies, most teams produced a functional prototype but had not refined the business model or dissemination plan. Although it can take years for the prototypes to become products ready for market, IDDS' connection to academic research institutions provided a means for work to continue beyond the conference. Since there was typically no funding earmarked to continue projects, IDDS helped participants raise funds, recruit new team members, form partnerships with implementers, or found new ventures. Some participants returned to future IDDS conferences with new project ideas or to further work on a previous IDDS project. IDDS conferences built a diverse global network of designers who supported one another in innovation and entrepreneurship.

Preliminary hypotheses and anecdotal evidence suggests that the intentional eclecticism of IDDS could be critical for innovation and transformative for participants and their communities. The democratic and participatory ethos of IDDS challenges societal hierarchies that typically hinder interactions between members of different social groups. For example, despite their limited formal education, artisans (e.g., welders, carpenters, mechanics, etc.) could demonstrate their ingenuity and teach fabrication skills to academics and professionals. Moreover, exposing people outside the realm of design, to the design process has the potential to expand their capabilities and change their view of their own self-efficacy and agency. For example, after an IDDS conference, a Tanzanian bicycle mechanic invented a solar-water heater and pedal-powered drill presses, blenders, and hacksaws. He and several other IDDS participants have also started design education programs and technology innovation centers in their communities. IDDS has been especially transformative for female participants since gender norms in some cultures associate technology with masculinity. In addition to women realizing their ability to create and use technologies, some IDDS technologies (e.g., grain threshers and mills) have the potential to shift the division of labor from manual labor by women to automated tasks for men.

3 Analysis of Project Case Studies

Twelve projects were completed in IDEO.org's 2011 Fellowship year. Fifty-six projects were completed by IDDS between 2007 and 2012. Four IDEO.org projects and five IDDS projects from this time period were analyzed based on various project features, and insights were drawn from this comparison. As an example of the project analyses, Table 1 compares two projects tackling the challenge of providing clean water to low-income communities.

Table 1 Comparison of design thinking and CCB as applied to two water projects

	IDEO.org - SmartLife	IDDS - Zimba
Project brief	Design a scalable *business* to sell water alongside nutrition and hygiene products to urban Kenyans	Design a *device* to automatically add the appropriate dose of chlorine to water as it flows out of hand pumps
Motivation	Create new sales channels and multiply health benefits for users by integrating water, nutrition, and hygiene	A personal mission of an Indian inventor to improve the lives of low-income people through innovative technologies
Design team	Architect, business strategist, and engineer led by an industrial designer	Engineering students led by a community member
User research	Semi-structured interviews with 28 customers, 13 entrepreneurs, and 1 government agency. A local market research firm in Nairobi, Kenya set up interviews based on user profiles	Inventor's prior experience installing hand pumps in villages provided insights for the context of use. The team visited villages to conduct informal tests, interviews, and observations
Prototyping	Translators role-played as SmartLife employees and sold branded water and health products to customers	Functional prototypes were made using fiberglass, simple hand tools, glue, and plastic sheets and tubes
Intermediary objects	Storyboards, brand identity, financial analysis spreadsheets	Physical prototypes, CAD (Computer-Aided Design) models, business plans

IDEO.org partnered with Unilever, Water and Sanitation for the Urban Poor (WSUP), and the Global Alliance for Improved Nutrition (GAIN) to design a scalable business selling clean water along with hygiene and nutrition products (Bigio et al. 2012). During two weeks of fieldwork in Nairobi, the design team conducted 50 interviews and set up a mock business and fake brand, selling 520 L of water in two days (Bigio et al. 2012). They tested several touch points in isolation and designed a business model including retail locations, a delivery service, and door-to-door sales representatives (Bigio et al. 2012). The resulting SmartLife brand highlights the convenience and reliability of the service rather than the traditional focus on health that most water initiatives emphasize. The design team proposed a pilot testing two concepts: Aspirational Wellness (drinking water with carefully curated personal care products) and Everyday Essentials (all-purpose water with familiar household and hygiene products) (Bigio et al. 2012). The first SmartLife kiosk opened in February 2013 in Kenya with plans to obtain 500 families as customers within their first next six to nine months.

At the 2009 IDDS conference in Ghana, an inventor from India proposed a project to design a device to automatically chlorinate water in villages. Chlorine is affordable, readily available, and effective for treating most waterborne pathogens, however the education and behavior change required to properly use chlorine for water treatment have been barriers to adoption. The IDDS Doser team designed and prototyped a device that could accept intermittent and variable water flow and dispense an appropriate amount of chlorine into the water. The Doser team recruited UC Berkeley and MIT students to continue the research and was awarded a $20,000 research grant from the National Collegiate Inventors and Innovators' Alliance. In 2013, Zimba was founded as a for-profit company with a patent pending for the chlorine doser. Zimba partnered with NGOs, research centers, and universities for pilots in India and Bangladesh. During a one-year pilot, 2 million liters of water were purified by two Zimba chlorine dosers at a cost of five rupees or $0.10 per 10,000 L of water.

Both SmartLife and Zimba reached new customers with new offerings providing safe water. For SmartLife, the business was the enabler, whereas for Zimba, technology was the enabler. SmartLife's innovation was a high-touch subscription service providing low-income customers with the reliable delivery of clean water and health products. Rather than inventing new technologies, SmartLife operated with existing technologies. In contrast, Zimba's innovation was a device that minimizes behavior change and offers robust functionality in resource-constrained environments. The Zimba doser made automatic chlorination affordable for individuals, households, and communities for diverse situations (e.g., chlorination at the point of collection, in homes, or at water kiosks and food stalls) without electricity or piped water.

Although it is beyond the scope of this paper to assess effectiveness or impact, some differences between SmartLife and Zimba are indicative of the different operational models of IDEO.org and IDDS. SmartLife was able to progress from concept to pilot in less than a year whereas Zimba moved from concept to pilot in about three years. However, SmartLife's budget was at least ten times more than Zimba's. SmartLife also had the advantage of several full-time salaried and experienced professionals working for established organizations with proven success. Zimba's staff consisted mostly of part-time student volunteers or recent graduates who had limited or no experience launching a product or business. The different trajectories of these projects might have been influenced by the differences between the operational models of professionals consulting for established organizations as compared to entrepreneurship and invention driven by budding innovators.

4 Discussion and Future Research

Both organizations are continually evolving. IDEO.org initially planned to train an entirely new cohort of Fellows each year rather than having design team members as permanent staff. However, in actuality, some designers from the Fellowship program continued working with IDEO.org as permanent staff. One of the 2011 Fellows, who worked for IDEO.org since its launch, became IDEO.org's Global Design Director in 2017. IDEO.org has since ended the Fellowship program in favor of hiring design professionals for permanent roles. IDEO.org's strength has been exemplary innovation and design to support national and multinational organizations in more effectively offering products, services, and programs all over the world. As of 2020, IDEO.org had design studios three locations: San Francisco, CA; New York, NY; and Nairobi, Kenya. To teach HCD to people all over the world, IDEO.org partnered with +Acumen to offer massive open online courses which has reached hundreds of thousands of people and a Design Toolkit that had been downloaded over 1 million times as of 2020.

IDDS has been continually evolving its model and curricula. When the conference location moved from MIT to Ghana in 2009, IDDS added lectures on Ghanaian culture and user research to prepare design teams to conduct interviews and observations with users in partner villages. IDDS also developed hands-on "Build-It modules," in which IDDS participants learned prototyping skills by making various technologies. For example, participants learned about basic electronics and hand tools by making a solar lantern. In 2010, IDDS shifted focus from creating early stage technologies to advancing prototypes to products and projects to ventures, and new lectures were created to teach business plan design, manufacturing at different scales, and supply chains management. In 2012, IDDS was held in Brazil, where lectures were bilingual with real-time translation between Portuguese and English. This was also the first IDDS to offer projects for urban areas in addition to the typical focus on projects for villages. In 2012, the International Development Innovation Network (IDIN) was formed to support IDDS alumni. In 2013, IDIN began teaching trainers and disseminating the IDDS approach so that institutions and groups all of the world could take the initiative to create and host conferences based on the IDDS curricula and model. From 2013 through 2019, IDIN supported various groups as they organized conferences focused on themes, such as maternal and neonatal health, rethinking humanitarian relief, and zero waste strategies. As of 2020, IDIN announced the postponement of all IDDS conferences due to the COVID-19 pandemic.

In summary, IDEO.org and IDDS had the same fundamental goal of design for poverty alleviation but there were significant differences in their approaches, as shown in Table 2. Aiming for measurable impact, IDEO.org tended to focus on products or services for partners who had already achieved some influence or scale. With the goal of building design capacity, IDDS inspired and taught people from a wide range of educational, occupational, or social statuses to invent technologies and launch ventures in villages and urban neighborhoods. The imperative for future research is not to evaluate which approach is better, but to determine what can be learned from each (Vechakul and Agogino 2016).

This paper has mainly focused on the similarities and differences between IDEO.org's design thinking and IDDS' CCB as transdisciplinary human-centered design (HCD) approaches for poverty alleviation. However, perhaps what are more pertinent are the emerging patterns of the influence of "HCD for poverty alleviation" on innovation, on development practitioners and users, and on design education.

Designing for the resource-constraints environments of low-income communities may expand the forefronts of innovation and generate solutions applicable for our over-consuming society. Constraints force designers to strive towards more *elegant* (cleverly simple and unusually effective) solutions that use appropriate resources more efficiently. For example, since the Zimba doser must function off-grid, the

Table 2 Comparison of IDEO.org and the International Development Design Summit

	IDEO.org	IDDS
Design approach	Design Thinking	Creative Capacity Building
Operation model	Partners pay a fee for service to IDEO.org for 6 to 12 week design consulting projects	Donors fund annual 3- to 5-week conferences teaching design with lectures, hand-on activities, and team projects
Mission	1.Increase the impact of partner organizations through design 2. Train social sector leaders in design thinking 3. Create resources to share HCD methods and processes online	1.Develop early-stage appropriate technologies 2. Create a global innovation network of entrepreneurs and inventors 3. Increase capacity for technology creation in developing countries
Design team	Professionals in industrial design, business strategy, engineering, journalism, and information design	Potential change agents (e.g., villagers, mechanics, students, teachers, doctors, farmers, masons, priests, and artists)
Scale	National or global	Villages and urban neighborhoods
Deliverable	Business models, brands, products, experiences, services, strategies	Appropriate technologies that can be operated and maintained locally
Strength	Emotional meaning	Elegant functionality
Strategies for impact	Storytelling and high-quality design inspire partners to implement	IDDS participants build prototypes, develop products, and launch ventures

Zimba doser works on gravity and does not require any fuel or electricity. Since hinges, levers, and valves often wear out and need replacement, the Zimba doser was designed with no moving parts to ensure robustness, especially in remote areas where parts and supplies are scarce. This simplicity minimizes costs and enables the Zimba doser to be produced and maintained with materials and manufacturing processes that are commonly available in developing countries. Moreover, the Zimba doser's ability to accurately and consistently measure and mix fluids may be applicable for purposes other than water treatment. Shawn Frayne—founder of Haddock invention and member of the IDDS network—refers to *confluent technologies* as innovations that emerge out of the scarcity and extreme pressures of low-resource areas to leapfrog over incremental or wasteful technologies in developed regions.

Transdisciplinary design incorporates some components of systems thinking to create holistic solutions that address the system rather than isolated aspects of the challenge. Transdisciplinary design extends the boundaries of the design space beyond focusing on economic or technological factors, to consider the socio-political, cultural, environmental, and ethical implications, thereby increasing the likelihood that a solution will be adopted and sustained effectively (Findeli 2001). For example, in designing the SmartLife business model, the IDEO.org design team considered both the "micro and macro elements of the entire ecosystem [...], including the customer experience, the business model, the financial breakdown, and the brand expression" (Ogbu 2012). The design team "constantly zoomed in and zoomed out, making sure that the pieces make sense both individually and working together" (Lidgus 2012).

Exposure to the design process as a structured way of framing a problem, generating innovative concepts, and refining and implementing a solution may empower development practitioners and low-income communities by encouraging people to try new ideas, experiment iteratively, and effect change. "CCB postulates that technology creation can be one pathway for an individual to identify or affirm their own abilities, to invite communities to seek solutions together, and to build towards meaningful influence over their lives and livelihoods" (Taha 2011). A preliminary evaluation of three-day CCB workshops conducted in Pader, Uganda suggests that CCB may encourage communities to work collaboratively "to develop technologies to meet their needs and/or generate income" and that individuals felt more "empowered to produce, repair, and adapt things" (Taha 2011). In fact, within a month of a CCB workshop in Pader, Uganda, community members had created 13 technologies, including a pedal-powered knife-sharpener, a wooden cart, and storage pots for evaporative cooling (Taha 2011).

Integrating transdisciplinary design into high school and undergraduate curricula could empower a new generation of design thinkers to address the complex societal challenges of the future. Teaching a general creative problem-solving approach that focuses on understanding people and their needs could provide a broadly applicable framework that promotes critical thinking and the integration of knowledge across multiple disciplines (Leblanc 2009). Project-based service learning courses have also been shown "to improve retention, student satisfaction, diversity, and student learning" (Dym et al. 2005).

As transdisciplinary human-centered design (HCD) gains prominence for innovating solutions for poverty alleviation, it will be increasingly important to conduct more in-depth studies on its utility for innovation, the adoption of resulting solutions, the impact on people who learn the design process, and implications for design education (Levine et al. 2016).

Acknowledgements The lead author is grateful to have worked as a Fellow IDEO.org and to have served as an organizer, participant, and team mentor for IDDS. We note that this study was conducted independently of these organizations and the findings reflect the authors' viewpoints, and not necessarily that of the organizations studied. This research was partially funded by NSF grant #1242232.

References

Arnstein, S. R.: A ladder of citizen participation. Journal of the American Institute of planners. 35, 216–224 (1969).

Boujut, J., Blanco, E.: Intermediary objects as a means to foster co-operation in engineering design. Computer Supported Cooperative Work. 12, 205–219 (2003).

Bigio, R., Ogbu, L., Friedberg, E., Vechakul, J., Shipp, J.: Water + Health in Kenya. Defining the Path to a Sustainable Business. IDEO.org, San Francisco, CA (2012).
Buchanan, R.: Wicked problems in design thinking. Design issues. 5–21 (1992).
Brown, T., Wyatt, J.: Design Thinking for Social Innovation. Stanford Social Innovation Review. 30–35 (2010).
Brown, T.: Change by design: how design thinking transforms organizations and inspires innovation. HarperBusiness (2009).
Brown, T.: Design thinking. Harvard Business Review. 86, 84–92 (2008).
Caruso, C., Frankel, L.: Everyday People: Enabling User Expertise in Socially Responsible Design. DRS (2010).
Das, S.: Automatic Chlorine Doser: Suprio Das at TEDxIIMShillong - YouTube, https://www.youtube.com/watch?v=-t9xPSlCRvc&noredirect=1 (2013).
Dym, C., Agogino, A., Eris, O., Frey, D., Leifer, L.: Engineering design thinking, teaching, and learning. Journal of Engineering Education. 94, 103–120 (2005).
Eisenbart, B., Blessing, L., Gericke, K.: Functional Modelling Perspectives Across Disciplines: A Literature Review. Proceedings of the 12th International Design Conference DESIGN 2012. 847–858 (2012).
Ertas, A., Maxwell, T., Rainey, V.P., Tanik, M.M.: Transformation of higher education: the transdisciplinary approach in engineering. Education, IEEE Transactions on. 46, 289–295 (2003).
Frayne, S.: Haddock Invention, https://www.haddockinvention.com/about/philosophy (2012).
Findeli, A.: Rethinking design education for the 21st century: Theoretical, methodological, and ethical discussion. Design issues. 17, 5–17 (2001).
Gasson, S.: Human-centered vs. user-centered approaches. Journal of Information Technology Theory and Application. 5, 29–46 (2003).
Hewens, S.: SmartLife is Open for Business Selling Pure Drinking Water, https://www.ideo.org/stories/smartlife-is-open-for-business-selling-pure-drinking-water (2012).
Leblanc, T.: Transdisciplinary Design Approach. Creativity and HCI: From Experience to Design in Education. 106–122 (2009).
Levine, D. I., A.M. Agogino, M. A. Lesniewski (2016). "Design for Impact: A Development Engineering Graduate Program at UC Berkeley," 32 (3B), Journal of Development Engineering, pp. 1396–1406.
Lidgus, S.: When the Brand is a Service, https://www.ideo.org/stories/when-the-brand-is-a-service (2012).
Ogbu, L.: When the Design Challenge is the System Itself, https://www.ideo.org/stories/when-the-design-challenge-is-the-system-itself (2012).
Polanyi, M.: The tacit dimension. University of Chicago press (2009).
Scholz, R. W.: Mutual learning as a basic principle of transdisciplinarity. Transdisciplinarity: Joint problem-solving among science, technology and society. Workbook II: Mutual learning sessions. 13–17 (2000).
Taha, K.A.: Creative Capacity Building in Post-Conflict Uganda, (2011).
Tietje, O., Scholz, R.W.: Embedded case study methods: Integrating quantitative and qualitative knowledge. Sage Publications, Incorporated (2001).
Vechakul, J., Agogino, A.: Human-Centered Design for Social Impact: Case Studies of IDEO.org and the International Development Design Summit. PhD diss. University of California, Berkeley (2016).
Wahl, D.C., Baxter, S.: The designer's role in facilitating sustainable solutions. Design Issues. 24, 72–83 (2008).

The Concept of Product Experience in Industrial Goods Development

Christian Wölfel, Sandra Olbrich, and Jens Krzywinski

Abstract The concept of *user experience* as a product development approach has been well established in interaction design. Today, it is also widely used in industrial design practice in the field of consumer products. However, human experiencing is important in the field of industrial goods as well. Due to differences in the whole lifecycle, experience approaches and methods cannot be transferred from consumer products to industrial goods without further ado. In this paper, we examine a theoretical framework of industrial goods experience and present first results of an empirical evaluation of this concept.

Keywords Product experience · User experience · Industrial goods

1 Introduction

Many design disciplines aim at bringing technology into line with humans. One core concept within these disciplines is *user-centered design* with a focus on needs of potential users. For a long period, the *usability* approach has been predominant. Its focus on effective fulfillment of tasks stands for technology as means of reaching well-defined goals. However, it is well known that products for private use are increasingly bought because of their potential of fulfilling needs that are not connected to rational goals but to prestige etc. Private users do not buy function but meaning (Enders and Hampel 2011). Fostered by psychological research on emotions, concepts on affective user reaction have been developed in the late 1990s (Hassenzahl and Tractinsky 2006). This finally led to a shift from *usability* to *user experience* and *product experience* in several design disciplines.

In the early 2000s, the industrial goods branch increased its interest in industrial design and human-centered product development. Despite its self-conception of being (almost) purely technology-driven, the importance of industrial design strategies has been proven for industrial goods as well (Herrmann and Möller 2009).

C. Wölfel (✉) · S. Olbrich · J. Krzywinski
Chair of Industrial Design Engineering, Technische Universität Dresden, Dresden, Germany
e-mail: christian.woelfel@tu-dresden.de

L. Blessing et al. (eds.), *The Future of Transdisciplinary Design*,
https://doi.org/10.1007/978-3-319-06382-9_15

Due to specific processes in this branch, the actual necessary shift to approaches of product experience has not been pursued yet. Another push for this process is coming from a new branding orientation of some major SME of industrial goods and more research projects about business-to-business branding in general. Since branding covers a broad range of disciplines it necessarily affects products experience as well as transdisciplinarity.

2 Theoretical Framework

2.1 Transdisciplinarity

Transdisciplinarity as a general concept of research has been described as the development of universal theoretical principles (Nicolescu 2002) or as integrated research (Mittelstrass 2007). According to Pohl et al., transdisciplinary research "addresses three kinds of research questions: (a) questions about the genesis and possible development of a problem field, and about interpretations of the problems in the life-world; (b) questions related to determining and explaining practice-oriented goals; and (c) questions that concern the development of pragmatic means (technologies, institutions, laws, norms etc.) as well as the possibility of transforming existing conditions" (Pohl and Hirsch Hadorn 2007). Mittelstrass (Mittelstrass 2007) argues that asymmetrical development of (practical) problems and (academic) disciplines is increasing due to the growing differentiation of disciplines. Accordingly, real problems increasingly demand for interdisciplinary and transdisciplinary research. According to Mittelstrass, transdisciplinarity goes beyond interdisciplinarity by developing science systems that are long-lasting and may change disciplinary rules. As Blevis & Stolterman (Blevis and Stolterman 2008) summarize Nicolescu's (Nicolescu 2002) concept of transdisciplinarity, as an "approach focusing on a broader goal, transcending disciplinarity and using collections of methods and their associated domains of expertise on an as needed basis as required". They conclude that this understanding applies to design research as well as design practice. This applies to many fields of design, but even beyond. As Ropohl argues, all technical sciences have never been purely disciplinary sciences but have to become aware of their transdisciplinarity and must still develop means to deal with it (Ropohl 2010). "Therefore, it is possible"—and necessary— "for design research to develop on its own needs and merits if more attention is paid to the notion of transdisciplinarity" (Blevis and Stolterman 2008). We agree on that, this paper is an example on transdisciplinarity across design disciplines as well as across broader academic disciplines as economics, psychology and design. A "problem in the life-world" (Mittelstrass 2007) is dealt with by transcending theories and methods across disciplinary borders without holding to disciplinary paradigms—but also without radically changing discipline.

2.2 Industrial Goods

Industrial goods are defined as "Goods that are destined to be sold primarily for use in producing other goods or rendering services as contrasted with goods destined to be sold primarily to the ultimate consumer.[…] The distinguishing characteristic of industrial goods is the purpose for which they are to be used, i.e., in carrying on business or industrial activities rather than for consumption by individual ultimate consumers or resale to them" (Bennett 1995). The distinction cited above impacts various aspects in the whole product lifecycle. In this paper, we will briefly discuss categories of industrial goods, their relation to stakeholders as well as specifics of the lifecycle process.

Categories. Industrial goods have been categorized according to different criteria. We refer to a sub-division by Geipel (1989). He uses the four criteria (*relations between subject and product, relations between product and environment, product complexity* and *relation between order and production*) to define groups of heavy and portable equipment. In the context of this paper, of these nine groups, *production systems, stand-alone production units, handling devices* and *commercial vehicles* are of interest.

Stakeholders. In contrast to consumer products, buyers are not the users in most of Geipel's industrial categories groups. There are differing perspectives and behaviors of buyers/decision makers, sellers, users, providers/operators and—if applicable—ultimate users. In many cases, those stakeholders are not single individuals. In most of the categories, multi-person scenarios are typical. That means there are not only different types of users (supervisors, operators, technicians, maintenance staff etc.), there are also multi-person development processes, multi-person buying processes and so forth. The individual experiencing of all stakeholders must be considered.

Development Process. In contrast to the consumer market, the industrial goods sector is characterized by e.g. derived demands, taylor-made solutions, non-standard prices (e.g. bidding), long decision phases, non-anonymous players or large quantities (Phadtare 2008). In the context of this paper, the most significant difference is the development process, where user-centered design as well as user experience hardly fits as it does in the early stages of consumer products development. Many industrial goods are not only made but also developed to order. Hence, in the industrial goods market (buying) decisions are based not on the product but rather on a trust in the supplier's ability to satisfy the buyer's (and user's etc.) needs (Hobday et al. 2000). Accordingly, many internal decisions must be made with strategy in mind. This will also apply to the approach how product experience is integrated into industrial goods development.

2.3 *Product Experience*

Human experiencing is understood as a—conscious and unconscious—ongoing reflection on events. It always incorporates the three dispositions *cognition, volition* and *emotion* (Hassenzahl 2008). Humans steadily experience objects and processes within their environment as a "constant stream of self-talk" (Hassenzahl 2008). The concepts of *user* (or *product*) *experience* focus on psychological effects elicited by the interaction between user and products (Hekkert et al. 2007). Although there are slightly different definitions and concepts of *user experience*, all refer to the fulfillment of various—more or less conscious—needs, concerns or values of users (Desmet et al. 2007; Hassenzahl et al. 2010). Most concepts focus on the affective reactions elicited by the interaction between human and product, particularly on emotions. However, in the context of industrial goods it is important to emphasize a concept of product experience that incorporates emotions as an essential disposition of any interaction between user and product. Particularly (consciously) emotional interactions are not excluded, but as a core it is subject to the concept of emotional design.

On the other side, product experience must be considered distinct from usability. Usability focusses on objective criteria of physiological and psychological ergonomics in order to provide products, which enable users to attain a goal in an easy and efficient way. However, usability can be a source of product experience by serving a concern (Desmet and Hekkert 2007). The concern of attaining a goal is one of the main dimensions of emotion eliciting described by the appraisal theory (Desmet and Hekkert 2007; Scherer 2001).

Within the product experience approach, most design methods target on defining user requirements. Most established methods are personas, narrative scenarios or contextual inquiry. In the domain of interaction design, methods for measuring product experience have been developed, e.g. the Positive Affect and Negative Affect Schedule *PANAS* (Watson et al. 1988) which is widely used. Another method of measuring user experience is the *AttrakDiff* questionnaire (Hassenzahl et al. 2003) which has been developed in the domain of interaction design. Due to its general applicability, it has been translated into different languages and has been incorporated in numerous research projects in different design domains in recent years (Hassenzahl et al. 2008).

The concept of product experience is a bold example of transdisciplinary research: there was a problem in practice and theory of (interaction) design which has been treated by the use of approaches, theories and methods from different academic disciplines. And beyond interdisciplinary research, the transfer across disciplines has changed design at least in the interaction design and industrial design domains. Within the research presented in this paper, the concept of product experience is again applied in another transdisciplinary research on industrial goods development. Figure 1 gives an overview of the approaches and theories of product experience/user experience that have been used as a basis for the theoretical framework of this research.

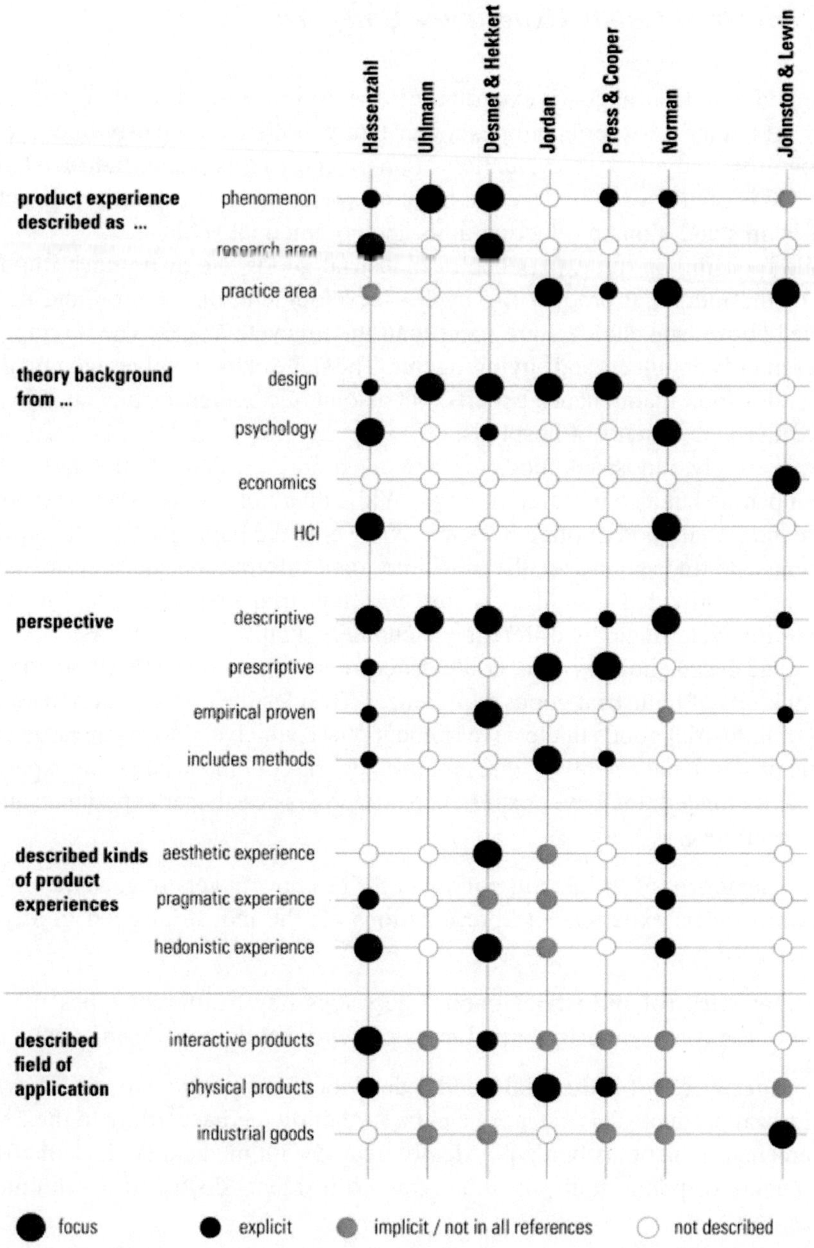

Fig. 1 Product experience as a transdisciplinary concept—theories and approaches from different disciplines

2.4 Industrial Goods Experience Categories

To begin with, industrial goods experience is the experiencing of an industrial goods product. We refer to experiencing as a constantly, holistic and partly unconscious, both cognitive and affective evaluation of an object. In this paper, industrial goods experience is mainly used to describe human experiencing of capital goods with an emphasis on stand-alone production units and commercial vehicles.

While focusing on the experience of industrial goods, we distinguish important stages of the industrial product life cycle—development, distribution and use. As explained above, stakeholders are specific to the lifecycle stages. The buying stage involves mainly engineers and buying-agents. The stage of use involves technical staff like operators and maintenance experts, but also public audiences like customers of manufacturers or patients of hospitals.

Aggravatingly, industrial goods are perceived on the customer side not only by different people but also in different ways. While operators get into physical contact with the industrial good, it often does not exist yet at the buying stage. Then printed documents and trade shows are the most important information sources, most information being provided visually. To sum up, industrial goods are experienced in different life cycle stage, by different stakeholders, in different contexts.

As stages, stakeholders and contexts correlate, we conceptually distinguish different kinds of industrial goods experiences. Besides the (anticipation of) experiencing of industrial goods in the development phase which already is an issue due to different professionals involved, three main categories of industrial goods experience remain to be treated: the buyer experience, the professional user experience and the public experience.

Buyer experience (of industrial goods) describes how buyers (e.g. as members of the buying center) experience (representations of) the industrial good in a typical pre-order setting.

User experience (of industrial goods) describes how (different types of) users experience the (final) industrial good in its professional domain during actual use.

Public experience (of industrial goods) describes how public persons experience the industrial product. This category can be further divided according to the level of involvement, e.g. patients being physically treated with medical devices in contrast to passengers of public transport in contrast to residents disturbed by commercial vehicles.

In theory, these categories are clearly divided and linked to specific stages, tasks, situations and persons. In reality, there can be overlaps between the categories, e.g. when identical stakeholders participate at multiple stages. Although brand, communication and services are important aspects of experiences in the industrial goods sector in general, we focus on the industrial products, e.g. the stand-alone production unit itself and its properties. Hence, the industrial goods buyer experience is referred to the industrial good at the stage of buying, not to the experience of the process. Table 1 gives an overview on how the three categories differ.

Table 1 Categories of industrial goods experience

	human	product	context	example	
buyer experience of industrial goods	recipient = buyer (engineer, purchaser) organizational decision process involving a group of people organizational and individual concerns homogeneous group of involved people	products do not exist mostly visual representations of similar machines at fairs, brochures evaluation of the whole product model (including image of producer)	work environment (professional experience) abstract consequences of buying decision competitors (bench mark, state of the art) political, legal, technological constraints	vendor acquisition on fair information procurement by brochures sales negotiation	affects buying behavior
user experience of industrial goods	recipient = user multiperson product usage, no possession individual within organizational concerns homogeneous group of involved people	usage in industrial context recitation and evaluation of the physical product based on appearance and function	industrial environment, work context (professional experience) machine environment depends on category (medical device, production unit, etc.)	operate a harvester gear cutting with CNC milling machine driving a bus	affects using behavior
public experience of industrial goods	recipient = private person no possession, multiperson product usage individual concerns inhomogeneous group of people	reception and evaluation of the physical product in service situation indirect product use	private environment (private experience) inhomogeneous machine environment and context of product recitation	patient in MRT walker watches harvester visiting a production plant driving in a bus or cab	affects (brand) image

3 Empirical Study

During the lifecycle of industrial goods, product experience is most relevant in the stages of presentation/sale, usage and coincidence with ultimate (end) users. Depending on the group of Geipel's sub-division (Geipel 1989), a different emphasis on buyers/decision makers, users, and end users must be made. That means while the experience-based assessment of industrial goods may not vary across the lifecycle stages for stand-alone production units or commercial vehicles, it may vary significantly e.g. for medical devices. Due to the specifics of industrial goods lifecycles, user experience must be integrated on a strategic level in general. However, it still has to find a place within single development processes.

The described approach and categorization of experience in the context of industrial goods has been transferred and derived from theoretical work from a number of disciplines. In the following steps this approach has to be applied and evaluated in the domain of design and design research. Most relevant questions at that stage focus on the evaluation of the theoretical framework as such as well as on consequences regarding design processes in the industrial goods domain.

As a pre-study, we conducted a series of semi-standardized interviews (n = 12) with experts from the industrial goods practice. The interviews have been conducted with purchasing, marketing and development experts of key accounts in the field of stand-alone production units. Within those 30 min interviews, the experts evaluated the experience of industrial goods in each of the three different categories: public experience, buying experience as well as user experience. The particular evaluation had been done using the *AttrakDiff2* questionnaire by Hassenzahl et al. (2003) as part of the interviews in addition to more general questions and depictions of industrial goods in the three categories. Within the *AttrakDiff*, there 28 items grouped to four independent dimensions *pragmatic quality* (PQ), *hedonic quality-identity* (HQ-I), *hedonic quality-stimulation* (HQ-S) and *attractiveness* (ATT). We conducted a univariate analysis of variances (ANOVA, F Hassenzahl and Tractinsky 2006; Bennett 1995) across the three categories of industrial goods experience for each of the four *AttrakDiff* dimensions. According to this, there is no significant difference between the product experience assessment of the same industrial good across the three categories *public experience*, *buying experience* and *user experience* (cf. Table 2). However, there are single items of the *AttrakDiff* evaluation that differ significantly between the categories of industrial goods experience. Interestingly, these are for instance the items *amateurish—professional* or *impractical—practical* that relate closely to actual usage. However it must be stated that these differences may result from the fact that some of the experts had issues with answering specific *AttrakDiff* items (as described in the qualitative comments analysis below). These difficulties varied across in particular categories of industrial goods experience and may be the primary explanation of the statistically proven differences.

In addition the *AttrakDiff* questionnaire items, the experts have been asked more general questions. Generally, all of the persons interviewed see themselves as experts in the field of industrial goods, most in the category of stand-alone production units (machine tool industry and plant engineering). The majority work as distributers (n = 7), all others as developers and application engineers. All of the test persons stated that they know the scenario of using a machine similar to the one shown in the depictions.

None of the test persons claimed to be an industrial design expert or has to deal with industrial design issues in their own work, likewise just one of them was familiar with the concepts of product experience or user experience. However, according to

Table 2 ANOVA of the four AttrakDiff dimensions *pragmatic quality* (PQ), *hedonic quality-identity* (HQ-I), *hedonic quality-stimulation* (HQ-S) and *attractiveness* (ATT) across the industrial goods experience categories *public* (PX), *buying* (BX) and *user experience* (UX)

Dimension, *item*	M_{PX}	M_{BX}	M_{UX}	F	*p*
PQ	0.38	0.39	0.25	0.497	0.61
HQ-I	0.75	0.80	0.37	1.120	0.34
FQ-S	−0.02	0.32	0.56	0.650	0.53
ATT	0.51	0.83	0.27	1.072	0.35
Professional	1.83	1.92	0.42	5.685	0.01
Practical	1.75	1.08	0.42	4.391	0.02

the answers, industrial design is an important factor for nearly all companies. Still, a large part of the test persons are unsure about the concrete, e.g. economic, outcome of industrial design. Accordingly, the importance of industrial design as part of product development has been rated diverse from very low to very high. Not surprisingly, user requirements are rated less important than technical or financial requirements.

To all of the test persons, user-oriented requirements analysis seems important. However, most test persons understand user-oriented as buyer-oriented. The buyer/purchaser is the most important source of information and requirements. Additionally, requirements like noise reduction are standardized and often taken from guidelines. Although qualitative analysis could not prove a difference of product experience between the public, buyer and user experience categories, nearly all test persons suspect varied assessments of the buyer and the user experience. The non-engineers among the test persons tended to evaluate the machine in a more holistic view (not just technical performance), but have no expectations concerning the stimulation (specific appearance). Some of the *AttrakDiff* items have been criticized as not matching the scenario, e.g. *gets me closer to people vs. separates me from people* was irritating especially in the public and buying categories.

4 Discussion

The pre-study has been conducted with a small sample size. We are currently increasing the number by conducting more expert interviews. Based on a larger sample, we will be able to investigate how far the results differ in the three selected industrial goods sectors *stand-alone production units, commercial vehicles*, and *medical devices*. Also, we will be able to clarify whether measured differences result from questionnaire items not matching the depicted scenarios or from actual differences of product experience. The answers will be important for further methodological research. If the experience of industrial goods is "the same" no matter if in the public, the buying or the user scenario, there is no need for specific attention or specific methods in the development process. Hence, established methods of user experience design could more straightforwardly be transferred to industrial goods development. Still, there is the specific lifecycle process of industrial goods which does not fit the consumer products processes. Hence, the way industrial goods experience is treated in design processes may not be transferred from the consumer goods field without further ado. Additional research is currently being conducted in the form of qualitative case study analyses on the integration of product experience methods in different industrial goods sectors. Based on the results, there may be the evolvution of a transdisciplinary approach of industrial goods design that incorporates methods and knowledge from different disciplines in order to deal with practical problems. If so, the development of a general culture of awareness towards the holistic experience of industrial goods will play a major role in this process. The culture itself needs to be established and balanced between and with many different departments. The engineering and marketing department will be in the focus during this process which has to include the management as well. Accordingly, branding is moving from mere aesthetic regulations over strategic approaches to content and narrative branding (Buck 2013). So all these activities foster each other by heading for a convincing industrial product experience of a unique company and brand.

This paper describes a form of transdisciplinary design research that is as transdisciplinary as any design research dealing with real-world problems should be—and usually is. In order to find an approach to deal with the research problem, theories, concepts and methods have been transferred from other disciplines "on an as needed basis" (Blevis and Stolterman 2008). Those kinds of transdisciplinary research, in this case applying concepts from economics, psychology and design to the field of industrial goods development, neither leads to new disciplines, nor replaces (parts of) the discipline. Once brought to a sufficient level, the findings may be relevant to the disciplines lending theories and methods. Such Examples of design research are a good argument for fostering transdisciplinarity—which is seen as a kind of repair to the current academic landscape which is characterized through increased disciplinary particularization, making disciplinary borders become borders of epistemology (Mittelstrass 2007).

References

Bennett, P.D.: Dictionary of marketing terms, 2nd ed. NTC Business Books, Lincolnwood, Ill., USA (1995)

Blevis, E., Stolterman, E.: The confluence of interaction Design & Design: from disciplinary to transdisciplinary perspectives. In: Undisciplined! Proceedings of Design Research Society Conference 2008, p. 344/9 (2008)

Buck, A.: Changing Branding. GRID graphic + industrial design 1(4), 33–39 (2013)

Desmet, P.M.: Product emotion. In: Schifferstein, H.N.J., Hekkert, P. (eds.) Product Experience, pp. 379–397. Elsevier Science, Oxfort (2007)

Desmet, P. M. A., Hekkert, P.: Framework of product experience. International Journal of Design 2007, 57–66 (2007)

Enders, G., Hampel, D.: System coding. Die Zeichen machen den Unterschied. Fruehwerk Verlag, Berlin (2011)

Geipel, P.: Industriedesign als Marktfaktor bei Investitionsgütern. PhD thesis, Ludwig-Maximilians-Universität Munich (1989)

Hassenzahl, M., Burmester, M., Koller, F.: AttrakDiff. Ein Fragebogen zur Messung wahrgenommener hedonischer und pragmatischer Gestalt. In: Szwillus, G., Ziegler, J. (eds.) Mensch & Computer 2003. Interaktion in Bewegung, pp. 187–196. B. G. Teubner, Stuttgart (2003)

Hassenzahl, M., Burmester, M., Koller, F.: Der User Experience (UX) auf der Spur. Zum Einsatz von www.attrakdiff.de. In: Brau, H., Diefenbach, S., Hassenzahl, M., Koller, F., Peissner, M., Röse, K. (eds.) Usability Professionals 2008, pp. 78–82. Fraunhofer, Stuttgart.

Hassenzahl, M., Tractinsky, N.: User experience. A research agenda. Behaviour & Information Technology 25(2), 91–97 (2006)

Herrmann, C., Möller, G.: Design governance. Design as a key factor for innovation and economic success. Icfai University Press, Hyderabad, India (2009)

Hassenzahl, M.: User experience (UX). Towards an experiential perspective on product quality. In: Brangier, É., Michel, G., Bastien, J. M. C., Carbonell, N. (eds.) Proceedings of the 20th

International Conference of the Association Francophone d'Interaction Homme-Machine on—IHM '08, pp. 11–15. ACM Press (2008)
Hassenzahl, M., Diefenbach, S., Göritz, A.: Needs, affect, and interactive products—Facets of user experience. Interacting with Computers 22(5), 353–362 (2010)
Hekkert, P., Schifferstein, H. N. J.: Introducing Product Experience. In: Schifferstein, H. N. J., Hekkert, P. (eds.) Product Experience, pp. 1–8. Elsevier Science, Oxford (2007)
Hobday, M., Rush, H., Tidd, J.: Innovation in complex products and system. Research Policy 29(7–8), 793–804 (2000)
Mittelstrass, J.: Methodische Transdisziplinarität. Lifis Online 1, https://www.leibniz-institut.de/archiv/mittelstrass_05_11_07.pdf (2007). Accessed 13 March 2013
Nicolescu, B.: Manifesto of transdisciplinarity. State University of New York Press, Albany; Great Britain (2002)
Pohl, C., Hirsch Hadorn, G.: Principles for designing transdisciplinary research. Oekom, Munich (2007)
Phadtare, M. T.: Industrial Marketing. Prentice-Hall of India Private Limited, New Delhi (2008)
Ropohl, G.: Jenseits der Disziplinen. Transdisziplinarität als neues Paradigma. Lifis Online 3, https://www.leibniz-institut.de/archiv/ropohl_21_03_10.pdf (2010). Accessed 13 March 2013
Scherer, K.R.: Appraisal considered as a process of multilevel sequential checking. In: Scherer, K.R., Schorr, A., Johnstone, T. (eds.) Appraisal Processes in Emotion. Theory, Methods, Research, Series in Affective Science, pp. 92–120. Oxford University Press, USA (2001)
Watson, D., Clark, L.A., Tellegen, A.: Development and validation of brief measure of positive and negative affect. The PANAS scales. Journal of Personality and Social Psychology 54(6), 1063–1070 (1988)

Team Diversity in Transdisciplinary Design

Overconstrained and Underconstrained Creativity: Changing the Rhetoric to Negotiate the Boundaries of Design

Claudia Eckert and Martin Stacey

Abstract What people think creativity is and what constitutes designing influences how designing is organized and carried out, and how design colleagues interact. In contrast to engineering, the fashion industry sees design as the process from idea to specification carried out by designers, and creativity only as open-ended and unconstrained. This reflects widespread beliefs about creativity and rhetoric about design. In knitwear, much detailed design is done by technicians converting these specifications into a program for knitting a garment. This often requires creative problem solving in finding a way to realise an idea or in optimising production without compromising the aesthetic appearance. Knitwear designers and technicians seldom co-design, but only a collaboration between designers and technicians can lead to an exploitation of the full potential of modern production machinery. This observation has implications for interactions between artistic and technical designers in a variety of other industries.

Keywords Design collaboration · Co-design · Creativity · Design process · Design management · Design education · Knitwear

1 Introduction: Requirements for Effective Collaboration

Many products transcend the narrow boundaries of disciplines, and all but the simplest design processes involve the collaboration of experts from different disciplinary backgrounds. How can people trained in specific disciplines collaborate effectively?

This paper contrasts two very different design processes, knitwear design and diesel engine design, to argue that successful collaboration across disciplines depends on a shared understanding of design itself, and an appreciation for other people's tasks

C. Eckert (✉)
School of Engineering & Innovation, The Open University, Milton Keynes, UK
e-mail: claudia.eckert@open.ac.uk

M. Stacey
School of Computer Science and Informatics, De Montfort University, Leicester, UK

L. Blessing et al. (eds.), *The Future of Transdisciplinary Design*,
https://doi.org/10.1007/978-3-319-06382-9_16

and skills. Also important is a reduction of the hegemony of one discipline over its "supporting" disciplines. For effective collaboration across disciplines, understanding the drivers that impel different design behavior is critical to assuring mutual respect. This paper advocates examining the nature of the constraints governing a design problem as a way to analyze different perspectives on design and enable colleagues with different perspectives to understand each other and collaborate more effectively.

The rhetoric of design is dominated by artistic design domains. This is reinforced by the public rhetoric around the "creative industries", which include many artistic design domains, such as fashion design and graphic design, but also broadcasting, advertising or antiques dealing, as the term was originally coined by Caves (2000) to cover domains with uneven and irregular remuneration. Therefore both engineering design and software design are excluded from this definition, which has affected the public perception of technical fields as uncreative.

2 Creativity in Design

A central part of creative thinking is the creation of novel mental structures through the combination of elements of different mental spaces (sets of objects and relationships). Creativity in design involves the coevolution of the designer's understanding of both the problem and the solution, which come together in a "creative event", which bridges the two mental spaces to produce an idea for a design that meets the needs of the problem (Dorst and Cross 2001). Fauconnier and Turner (2002) argue that bisociation, the integration of two previously separate mental spaces (Koestler 1964), is central to thinking, not just to exceptional creative acts. New design ideas involve new mental representations of both the design and its context (Visser 2006). In most design work, the production of new ideas is tightly coupled to the development of sketches or other external representations of the design that serve as triggers for both reinterpretation and extension as well as external memories (Schön 1983). Research on analogical reasoning and conceptual combination shows that the construction of coherent mental representations is guided by the satisfaction of multiple constraints imposed by the different elements on each other and by the purpose driving the conceptual combination (Holyoak and Thagard 1995; Thagard 1989, 2000).

Boden (1990,1999) defines creativity as the generation of ideas that cannot be produced by the same set of generative rules as other, familiar ideas. She differentiates between two forms of creative thinking: exploratory creativity which applies existing mental operations and elaborates existing ideas or concepts; and transformational creativity, which alters the conceptual framework to make previously impossible thoughts possible. The quality of creative ideas is typically assessed in terms of their novelty and appropriateness (Massetti 1996) and unobviousness (Howard et al. 2011); transformational creativity produces less obvious results but is rare.

Design problem spaces are different between artistic domains, such as knitwear design, where the designers narrow the design space by making a series of decisions that reduces their search space to manageable proportions; and technical domains, where problems are often overconstrained and designers have to find ways of loosening constraints to reach a well-constrained problem. Reframing problems to perceive and reason about the constraints in different terms is often a crucial part of both kinds of process (Stacey and Eckert 2010); however in artistic design processes creativity is often equated with novelty, while in technical domains creativity often lies in finding a solution that is similar to an existing design, but meets all the constraints (Eckert et al. 2012). In some situations, a novel formulation of the problem, or characterization of the space of acceptable designs, may be the crucial creative step, while constructing a design can be relatively straightforward (Eckert and Stacey 2001).

3 Transdisciplinary Design

Our understanding of academic and professional disciplines has evolved for centuries and has been institutionalized through the establishment of university departments and formal professional training. In design and engineering this largely took place in the nineteenth century. Much of this disciplinary training is geared to solving the problems that designers in industry actually meet. However, many modern products don't fit into single disciplines, and require collaboration across expertise boundaries.

Transdisciplinary approaches to researching and addressing social and environmental problems are becoming increasingly fashionable. The term 'transdisciplinary' has been defined and used in different ways, sometimes as a synonym for the more established concepts 'multidisciplinary' and 'interdisciplinary'. But the distinctions between 'multidisciplinary', 'interdisciplinary' and 'transdisciplinary' are crucial for understanding what's going on (or what should happen) in collaborative activity crossing traditional disciplinary boundaries. Here are the conceptualizations we use.

Multidisciplinary activity is the application of the concepts and methods of different disciplines to a common problem, in a more-or-less coordinated way. *Interdisciplinary* activity is the application of the concepts and methods of one discipline to a problem posed and framed by another discipline: a tighter coupling of concepts and paradigms than is suggested by the term 'multidisciplinary'. By contrast, *transdisciplinary* activity blurs and transcends traditional disciplinary boundaries to achieve a more fluid synthesis, or a set of problem formulations, concepts and methods that defy conventional disciplinary classification (Nicolescu 2008). This, of course, is an easier thing to pay lip service to than to achieve. For good reasons, much design methodology and management practice is geared to achieving smooth handovers between tasks for specialists. How, then, could design practice grounded in the application of established concepts and techniques become transdisciplinary, involving a fluent melding of discipline-specific design problem formulations, concepts and methods?

One possibility is training people in more than one domain, so they can integrate *multiple disciplines in a single brain* and communicate with people from different disciplines and broker between them. For complex products this would require a breadth and depth of technical expertise that few have the opportunity to acquire, and still raises issues of division of labour. Engineers are trained to have T-shaped expertise, deep in their specialism but spanning a large part of engineering in less detail; however this won't be enough when projects involve non-engineers.

A disciplined *application focus* could use domain concepts to drive discussion and structure activities; however much design effort has to be focused on single discipline problems thus undermining true integration. Many writers see transdisciplinary design in terms of the co-option and integration of disciplinary contributions to real world problems on an as-needed basis.

Fluid collaboration involves multidisciplinary groups with participants with different disciplinary expertise, who fluently switch perspective to integrate the concerns and constraints of different disciplines.

Our concern in this paper is how effective fluid collaboration can be put in place, rather than whether or not collaborative design in practice is multidisciplinary or qualifies as transdisciplinary. In order for fluid collaboration to be possible, the participants in the design process need to understand each other's concepts well enough to exchange ideas and see what they need to do to formulate problems for discipline-specific designing, which requires a certain degree of shared or overlapping understanding. Shared understanding between diverse groups can be achieved through the shared activity of designing (Arias et al. 2000); this can be facilitated by boundary objects conveying information between different disciplinary specialists (Star and Griesemer 1989). The tension between different groups can also be one of the engines of design creativity (Fischer 2000). Coordinating design activities by handing over information with little or no interaction (as in most knitwear design) does not facilitate a convergence of views.

4 A Comparative Approach to Studying Design

This paper draws on comparisons between interview and observation based studies of designing in different industries over nearly 20 years, including several of large-scale engineering in the aircraft and automotive industries. Our research (Blackwell et al. 2009; Eckert et al. 2010) aims to identify the causes of both similarities and differences between design domains and therefore the transferability of skills and techniques across different domains.

Since 2002 we have studied diesel design in a UK manufacturer of off-highway diesel engines from the perspective of engineering change (Jarratt et al. 2005), process planning (Flanagan et al. 2007) and testing (Tahera et al. 2012).

We carried out the majority of our study of the knitwear industry in the 1990s when we studied communication between designers and technicians (Eckert 2001), as well as how the designers' creative thinking works in a study on the use of sources of inspiration in the knitwear industry (Eckert and Stacey 2001, 2000, 2003). More recently we went back to studying processes in the textile industry in the context of understanding how to make garment consumption more sustainable (Black et al. 2009). In 2013 we conducted two interviews with a knitwear technician currently working in the UK, who critically commented on our findings from the 1990s, when he started out, and how the industry has changed.

5 Collaboration, Understanding and Respect

Design is a social process which inevitably requires collaboration for all but the simplest products. We usually think of all participants as designers, who share an understanding of design in the same way. Looking closer this is by no means the case. Throughout the design process the skills of people we would not think of as designers are required. Engineers receive product briefs from marketing and other customer facing functions; they are often managed by people with a business background. They draw on findings from scientists. In engineering companies there is usually a consensus that the people who are engaged in defining the products are designers albeit with different flavours. For example in the design of diesel engines mechanical engineers work with electrical engineers, control engineers, computer scientists, etc. The company culture is dominated by mechanical engineering for historical reasons, but the other disciplines are valued for their expertise.

The knitwear design process forms a stark contrast in this respect. The process is primarily shared between knitwear designers and knitwear technicians. Even when they are co-located, they have little informal interaction and collaboration does not work well. The ambiguity of the representations through which design intent can be communicated by the designers forces the technicians to interpret what they receive based on their own experiences (Eckert 2001). This paper however focuses on one aspect of this badly working collaboration: the lack of understanding and to some extent respect the two groups have for each other. The designers see the work of the technicians as a type of translation or implementation and don't have the technical understanding to see how challenging it can be to create particular knitted structures or effects. They do not recognize the technicians' job as a creative design process in its own right. Technicians also often dismiss designs out of hand as unworkable, but when pushed deliver entirely satisfactory designs. Technicians on the other hand are not too impressed by the technical understanding of designers when they have to prove that something is unworkable. These problems are aggravated when sampling is moved offshore to suppliers. Designers and technicians cannot communicate directly and have little time to go through iterations. The effect is that the knitted garments in the shops don't use the enormous potential that is afforded by modern knitting machine technology. Only a few technicians like our informant, who works closely with the knitting machine makers, have the skills to exploit the machines' potential.

Knitwear designers are rarely fully aware of what the machine can do, because technical understanding is required to see the potential they afford. The technicians who have the skills are rarely give the opportunity to explore what the machines can do, but work through design specifications assigned to them. Their time is largely used to generate specific garments or carry out other tasks assigned to them by the designers. Therefore the innovations by technicians are in response to design requests, without designers being aware that they are asking for innovative design. Knitwear is also in engineering terms a non-linear problem, in that small changes to the visual appearance of a design can require very significant changes in how it can be achieved.

6 How Constraints Shape Design Behaviour

We have made the case elsewhere that design behaviour and the nature of the creativity it involves is primarily shaped by the nature of the constraints on the design (Stacey and Eckert 2010).

Constraints on design come from three different mutually-influencing sources: (a) The *problem* that the design must solve or the need that the design must meet; this includes product requirements, manufacturing requirements, and constraints stemming from the strategic goals of the company. (b) The *process* by which this is achieved. (c) The *emerging solution* – since making certain decisions will rule out or restrict options for other later decisions.

Constraints can vary along a number of dimensions in how they are expressed, perceived and assessed, which characterise the design process.

- *Explicit/implicit/tacit*. Explicit constraints are stated in the product brief or external sources that obviously apply to the design project, such as regulations or company guidelines. Implicit constraints can be inferred from the problem or its context but are not necessarily known by the people whom they affect. Designers can perceptually recognize violations of tacit constraints, and can learn to avoid them intuitively, without being able to articulate what the constraints actually are.
- *Internal/external* to the design project. Many constraints arise directly from the product itself and the process by which it is generated, however others are external to the immediate sphere of influence of the designers, like market factors, general fashion or economic uncertainties.
- *Objective/subjective* procedures to assess whether constraints are met. Many constraints or requirements on a product can be objectively assessed; in particular numerical constraints, like dimensions, strength, weight or manufacturing time can be assessed objectively, so that it is clear whether the constraint has been met. Others are highly subjective, like market appeal or fashionableness. Here the understanding and expertise of the designer is the main guide.

Design processes involve a combination of underconstrained problems where designers seek out and construct constraints, and problems which have difficult combinations of constraints even if they are not actually overconstrained. The balance between overconstrained and unconstrained aspects of design processes varies between domains and individual design problems. In generating a product, the designers need to resolve overconstrained aspects of the design into a clear specification that they can meet. They also need a way of identifying and deciding on a design solution for underconstrained aspects of the designs. Therefore designers often have to engage in constraint seeking behavior for underconstrained parts while activity reducing the constraints on other parts.

In artistic design processes, like knitwear design, very few explicit constraints exist for the product itself, beyond basic fit and material safety. It is critical to position the garment correctly in the context of other designs. A design must appeal to the target market and be appropriate in the evolving context of fashion. However, knitwear designs typically have very tight cost constraints, which are a mixture of material cost, knitting time, machine costs and assembly cost. The challenge for the knitwear designers is to narrow the design space in a suitable way. They study the context of the emerging fashion and first select themes and then materials with which they can design garments that suit these themes. They identify sources of inspiration both for the overall appearance of the garments and the detailed features, and generate sketches of the designs that they hand over to the technicians to turn into garments. The technicians need to realize the designs in a cost-effective fashion. As the materials are typically already selected, they have to optimize the knitting time to reach the right cost point. In terms of the constraints the knitwear designers and technicians address very different types of design problems. The knitwear designers engage in a constraint seeking activity, where many of the constraints are inherently implicit, external and subjective to assess, whereas the technicians have to implement designs to clear price points and quality criteria. Their design space has explicit, internal and objective constraints.

Engineering design, for instance diesel engine design, is very strongly constrained from the beginning of the process in terms of target requirements, performance constraints, product reuse targets, manufacturing processes etc. In fact the design problems are typically at least in part overconstrained with targets that are in direct conflict with each other; for example emission targets and space targets for tier 4 engines were in direct conflict as manufacturers had to accommodate large filters.

7 Contrasting Design Processes Seen in Terms of Constraints

Our two examples contrast significantly in terms of the constraints that are placed on both the design process and the product.

The emissions of diesel engines are tightly regulated. The constraints drive both the process in terms of delivery time, but also the product in terms of key performance parameters. Meeting emission targets drives the design process. The process starts with performance models which identify key requirements on the product and its components. The company takes a so-called requirement cascade approach (see

Wyatt et al. 2009), see Fig. 1, where they are introducing requirements incrementally into the design process. They start with key customer requirements for the performance and use of the product. The company sets novelty targets defining the amount of reuse of existing components and solution principles. These are reflected as key constraints on parameters in the mathematical models of the engine they are using. An FMEA gives them problem areas that they need to consider in the product, again reflected as parameter constraints. Emission targets are also set as constraints. In the next round the company concentrates on key performance parameters and considers conflicts between these parameters in the engine, as well as further customer parameters. At this point they are fairly sure that it is possible to design an engine with the desired characteristics, so that they can focus on durability and reliability, modifying geometry and materials requirements. They also consider business requirements such as manufacturability and the supply chain. After this the requirements cascade continues through the specifications for the product's components where the details of the components are designed. This is a somewhat iterative process as some component design has to take place to constrain the higher level performance modelling.

Most of these constraints are available at the beginning of the design process and typically pose an overconstrained problem. The company deals with this by prioritizing the constraints and dealing with them in batches. The constraints on diesel engines are largely explicit and the engine can be tested for them either in physical or virtual tests. Some constraints like engine noise are to some extent tacit

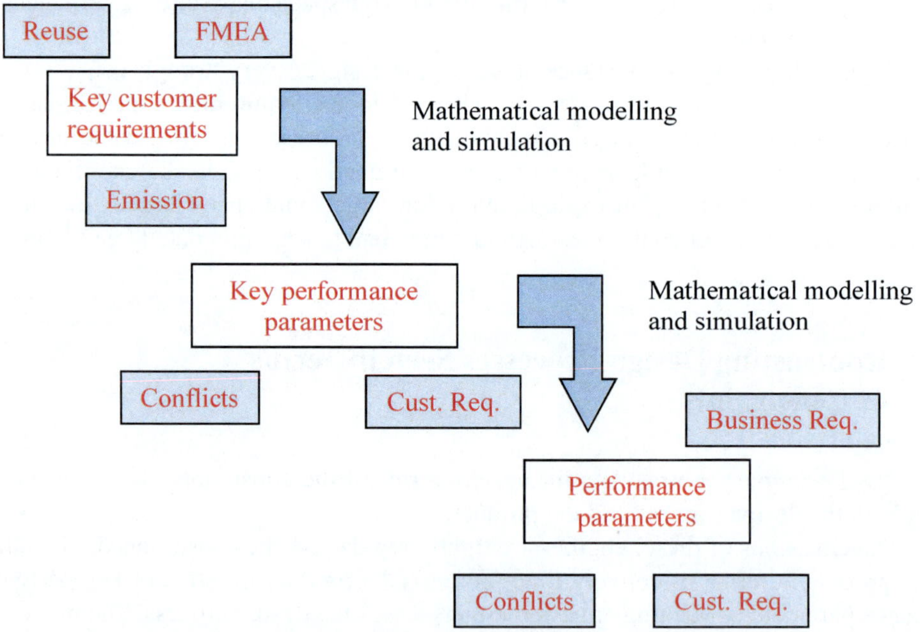

Fig. 1 Diesel engine requirement cascade

and subjective. The company has specialists who handle engine noise, who attempt to analyze and quantify the noise. The constraints arise from different disciplinary fields, but are handled in bundles. The mathematical modelling integrates different disciplinary perspectives in the models. By handling the constraints in groups the company generates underconstrained design problems in aspects of the product that have not yet been specified, which provides them with the possibility to think in an open-ended manner about their problems.

The profile of constraints looks very different in the case of the knitwear design process. The task of the knitwear designers is to design the look and feel of garments in a way that suits both their customers and evolving fashion. Designers develop an intuitive understanding of their target customers, and often picture personas for whom they design particular garments. However target customers are not explicitly specified and the boundaries between exciting and outrageous and between classical and dull are fluid and implicit. Similarly the context of fashion is only specified explicitly on a sketchy and impressionistic level. The space of acceptable garments within particular fashions is typically expressed with references to other garments without ever making explicit what aspects of the designs are carried over (Eckert and Stacey 2000). Much of the skill of a knitwear designer lies in perceiving and understanding this evolving context of fashion. They develop intuitions for the new designs and their intuitions change with seasons. Knitwear designers immerse themselves in pictures of designs and knitted garments, that they use as sources of inspiration, and adapt these objects for their new designs. Their creativity lies often in identifying and adapting a suitable starting design (Eckert and Stacey 2003). While they have a brief at the beginning of work on a season and a high level plan for the types of garments they want to design, these are fluid starting points rather than an objective set of criteria against which the final products can be assessed. The designs are evaluated perceptually. Even if the design is not what the designers originally had in mind, if it works designers make use of it. As they see new garments entering the market from competitors or leaders in their market segment, the designers add to their ranges and adapt their designs.

When the knitwear designers hand over to the technicians this rich tacit context is not shared. The technicians receive what looks to them a contradictory and ambiguous description of a design combined with tight targets in terms of overall costs and knitting time. Since knitting is composed of discrete stitches, the technicians need to find ways of fitting the knitted structures or patterns defined by the designers into a given number of stitches and rows according to the shape for the garment. Fitting the pattern onto a given shape while minimizing the knitting time is extremely fiddly. Modern knitting machines have CAD systems that work much of this out automatically, but understanding how the design can be tweaked to improve performance is still very much part of the core expertise of the technicians, who also have to take care to adapt the automatic program to the characteristics of the yarn. Programming a knitting machine requires meeting a number of explicit constraints that often conflict and finding clever ways to solve problems without changing design intent. If technicians run into problems they have to think of clever ways of achieving a similar visual effect with a different structure. As we argue elsewhere (Stacey and Eckert 2010; Eckert et al. 2012) this is very similar to creativity as it is required in engineering. Much of the conflict between designers and technicians we mentioned in Sect. 5 comes from miscommunication of design intent and thus how much and in which ways the technicians' designing is constrained by the specifications they get.

8 Conceptions of Design and Creativity as Barriers to Transdisciplinary Collaboration

The engineering designers did not talk in terms of creativity at all during our case study, but they expressed a great respect for the importance of the work of their colleagues, as well as the need to work on further integration through joint models and joint processes.

Different kinds of designing involve different forms of active constraint management, in which designers try to formulate the right problem to solve. This view of creative designing, corresponding to what design researchers like ourselves see designers doing, clashes with beliefs about creativity widely held by designers. Knitwear designers see themselves as creative, not because of what they do in detail, but because they are knitwear designers, thus part of the creative industries. In practice they follow a repeatable process that is shared in its fundamental structure by all knitwear design companies (Eckert 2006). They have clear heuristics and rules with which they can generate new design elements. Much of the skill of a knitwear designer lies in understanding the context of other products on the market and emerging trends and applying this to their target market to both meet the stylistic requirements of their customers and have enough differentiation from other products to draw customers in. By Boden's definition of a creative solution being one that could not be generated with existing rules (Boden 1990,1999), knitwear design would rarely be creative. In fact innovation in fashion products is very rare in practice.

For the knitwear designers, designing is an open-ended underconstrained process of finding a design that meets the constraints that they do have while fitting in the wider fashion and market context. Many of the requirements and the skills needed to meet them are tacit. While the designers realize that the technical realisation of a garment can be challenging, for them the design process ends with a technical sketch: in engineering terms a vague specification of a garment. The engineering equivalent would be to think of a design process as ending with concept selection. The technicians carry out the detailed design of the garment from concept to manufacturing instruction. To them much of what the designers do is travelling or looking at pictures, and while they realise it is necessary it does not quite seem like work to them. Both groups have different notions of design and don't acknowledge the challenges of the other as creative design activities. Knitwear technicians don't dwell much on their creativity, but are enormously proud of clever new solutions that they have generated. They see their work often as getting the knitting machine to do what they want it to do and enjoy getting the machine to do designs they have not knitted before.

A mutual recognition of each other's activities as design activities under different patterns of constraints might be a first step towards recognizing each other as equal partners in a co-designing process. For example companies need to recognise that a technician spending time on exploring the technical potential of a machine can be just as valid as a designer exploring upcoming fashions.

9 Conclusion

Effective co-design benefits from the participants in the design process understanding what the other participants contribute, and respecting it. The knitwear designers we have met generally respect what the technicians do as highly skilled work, but don't recognize it as being creative or as designing. The reason for this is a failure of design education: people graduate from design degrees as well as on-the-job professional training with little understanding of the nature of creativity, and thus little understanding of what their colleagues do that is creative. We found this to be worse among managers with the power to change things, who frequently had no design training. The consequence of this is not just front-loading the underconstrained, constraint-seeking, aspects of knitwear design. It is inefficient and ineffective processes: a lot of wasted time, and many situations where designers and technicians need each other's knowledge (to set or relax the constraints on their own problems), and either don't get it or only get it late and in an inefficient manner. Few knitwear companies prioritize enabling designers and technicians to talk, but those we have seen that do benefit from it.

The result is the production of products that could be better. Conceptualizing knitwear as a transdisciplinary is not necessary in the sense that a lot of very nice knitwear is being produced, but it might lead to smoother and more efficient processes and ultimately far more exciting clothes. Even in diesel engine design a more transdisciplinary mindset could bring an earlier integration of different experience and fewer iterations resolving small problems.

This has implications for many other industries where artistic designers work with more technically-oriented colleagues, such as car stylists and automotive engineers, and engineers and product designers. Transdisciplinary working gains from understanding where the other parties make decisions, when they need imagination and how they seek out or avoid constraints – thus how designers with different expertise can contribute to each others' constraint-setting activities. Respect for each other's contribution and its importance for the whole product is an important prerequisite to achieve transdisciplinary collaboration. This crucially depends on seeing each other as participants in creative designing, solving both overconstrained and underconstrained problems, which in turn depends on having a more sophisticated view of creativity, in conflict with the myths and rhetoric of creativity prevalent in the creative industries and society at large. A key step towards this is understanding the role of constraints in creative thinking (well understood by psychologists and design researchers) and the importance of constraint-seeking and constraint-relaxing in framing design problems. This is a challenge for design education.

Acknowledgements Our understanding of transdisciplinary activity has benefited from conversations with Professor Sue Thomas of the Institute of Creative Technologies at De Montfort University and the other habitués of the DMU Transdisciplinary Common Room.

References

Arias, E. G., Eden, H., Fischer, G., Gorman, A., Scharff, E.: Transcending the individual human mind: Creating shared understanding through collaborative design. ACM Transactions on Computer-Human Interaction 7, 84–113 (2000)

Blackwell, A. D., Eckert, C. M. Bucciarelli, L. L., Earl, C. F.: Witnesses to Design: A Phenomenology of Comparative Design. Design Issues 25(1), 36–47 (2009)

Black, S., Eckert, C. M., Watkins, P., Delamore, P., Geesin, F., Harkin, S.: Considerate Design for Personalized Fashion: towards sustainable production. Proceedings of MCPC09. Helsinki (2009)

Boden, M. A.: The Creative Mind: Myths and Mechanisms. Abacus, London (1990)

Boden, M. A.: Computer Models of Creativity. In: R.J. Sternberg (ed.) Handbook of Creativity. Cambridge University Press, Cambridge, UK (1999)

Caves, R. E.: Creative Industries: Contracts between Art and Commerce. Harvard University Press, Boston, MA (2000)

Dorst, K., Cross, N. G.: Creativity in the design process: co-evolution of problem–solution. Design Studies 22, 425–437 (2001)

Eckert, C. M.: The Communication Bottleneck in Knitwear Design: Analysis and Computing Solutions. Computer Supported Cooperative Work 10, 29–74 (2001)

Eckert, C. M.: Generic and specific process models: Lessons from modelling the knitwear design process. In: Proceedings of TMCE 2006, Volume 2, pp. 681–692. Delft University of Technology and University of Ljubljana, Ljubljana, Slovenia (2006)

Eckert, C. M, Blackwell, A. D., Bucciarelli, L. L., Earl, C. F.: Shared Conversations Across Design. Design Issues, 26(3), 27–39 (2010)

Eckert, C. M., Stacey, M. K.: Sources of Inspiration: A Language of Design. Design Studies, 21, 523–538 (2000)

Eckert, C. M., Stacey, M. K.: Designing in the Context of Fashion—Designing the Fashion Context. In: Designing in Context: Proceedings of the 5th Design Thinking Research Symposium, pp. 113–129. Delft University Press, Delft, Netherlands (2001)

Eckert, C. M., Stacey, M. K.: Sources of inspiration in industrial practice: The case of knitwear design. Journal of Design Research, 3 (2003)

Eckert, C. M, Stacey, M. K., Garthwaite, P., Wyatt, D. F.: Change as little as possible: Creativity in design by modification. Journal of Engineering Design 23, 337–360 (2012)

Fauconnier, G., Turner, M.: The Way We Think. Basic Books, New York (2002)

Flanagan, T., Eckert, C. M., Clarkson, P. J.: Externalising tacit overview knowledge: A model-based approach to supporting design teams. Artificial Intelligence for Engineering Design, Analysis and Manufacturing 21, 227–242 (2007)

Fischer, G.: Social Creativity, Symmetry of Ignorance and Meta-Design. Knowledge-Based Systems 13, 527–537 (2000)

Holyoak, K. J., Thagard, P.: Mental Leaps. MIT Press, Cambridge, MA (1995)

Howard, T. J., Culley, S., Dekoninck, E. A.: Reuse of ideas and concepts for creative stimuli in engineering design. *Journal of Engineering Design* 22, 565–581 (2011)

Jarratt, T. A. W., Eckert, C. M., Clarkson, P. J.: Pitfalls of Engineering Change: Change Practice during Complex Product Design. In: H. ElMaraghy & W. ElMaraghy (eds.) Advances in Design, pp. 413–423. Springer, London (2005)

Koestler, A.: The Act of Creation. Hutchinson, London (1964)

Massetti, B.: An empirical examination of the value of creative support systems on idea generation. *MIS Quarterly* 22(1), 83–97 (1996)

Nicolescu, B. (ed.): Transdisciplinarity: Theory and Practice. Hampton Press, Cresskill, NY (2008)

Schön, D. A.: The Reflective Practitioner. Basic Books, New York (1983)

Stacey, M. K., Eckert, C. M.: Reshaping the box: creative designing as constraint management'. Int. J. Product Development 11, 241–255 (2010)

Star S. L., Griesemer J.: Institutional ecology, 'Translations' and Boundary objects: amateurs and professionals on Berkeley's museum of vertebrate zoology. Social Studies of Science 19, 387–420 (1989)

Tahera, K., Earl, C. F., Eckert, C. M.: The Role of Testing in the Engineering Product Development Process. In: Proceedings of TMCE 2012. Karlsruhe, Germany (2012)

Thagard, P.: Explanatory coherence. Behavioral and Brain Sciences 12, 435–467 (1989)

Thagard, P.: Coherence in thought and action. MIT Press, Cambridge, MA (2000)

Visser, W.: The Cognitive Artifacts of Designing. Routledge, New York (2006)

Wyatt, D. F., Wynn, D. C., Clarkson, P. J.: A computational method to support product architecture design. In: Proceedings of the 2009 ASME International Mechanical Engineering Congress and Exposition, IMECE2009–11138. ASME, Buena Vista, FL (2009)

Project-Based Design and Transdisciplinarity: Rethinking Approaches to Spatial Design Education

Carmella Jacoby-Volk and Shoshi Bar-Eli

Abstract This paper proposes new approaches to spatial design education that are based on the synergy between the methodology of transdisciplinarity and Project Based Learning (PBL). The paper demonstrates that transdisciplinary methodologies remain theoretical and therefore cannot be implemented to solve complex real-life problems. Only by introducing risk-taking PBL methodologies, generating organic leadership and promoting short- and long-term learning using the components of transdisciplinary performance can real-life design projects be initiated to solve problems and empower all involved stakeholders. The paper analyzes PBL and transdisciplinary research (TR) methodologies as implemented in SpeeDesign projects. The projects were carried out in the Designers Clinic run by the Interior Design Department of COMAS (College of Management, Academic Studies). This clinic provides a unique platform positioned between design practice and academia with the aim of merging the field of design with the world of social involvement. The analysis demonstrates how the use of SpeeDesign can transform a spatial design engagement project into a new platform for emerging design methodologies. The analysis also shows how collaborative and participatory spatial design projects such as SpeeDesign can change spatial design education curricula by removing the current boundaries between academia and practice. To this end, all spatial design disciplines and practitioners—architecture, interior architecture, design and urban design practices, artists, social scientists and communities, as well as politicians, builders, entrepreneurs, economists, policy-makers and lawyers—must engage in reflection and debate toward establishing and implementing this type of short and long-term learning approach. Real-life projects can become a force for innovation and can generate change in the perceptions of all agents involved, as well as the overall perceptions, content, methodologies and outcomes of the entire discipline.

Keywords Transdisciplinarity · Participatory spatial design · Real-life project · Project-based learning · SpeeDesign · Curricula network

C. Jacoby-Volk (✉) · S. Bar-Eli
COMAS (Collage of Management Academic Studies), Rishon Lezion, Israel
e-mail: jacarmella@colman.ac.il

L. Blessing et al. (eds.), *The Future of Transdisciplinary Design*,
https://doi.org/10.1007/978-3-319-06382-9_17

1 Introduction

Design "is never a process that begins from scratch, to design is always to redesign" (Latour 2008, 3). The notion of redesign signifies the unique position of spatial design as a discipline related to the scale of the body and grounded in tactile and sensual experience, thus allowing for the assimilation of a participatory dynamic that redefines spatial design as a new force for active engagement. Redesign allows for the emergence of systems that emphasize fluidity, exchangeability and multiple functionalities—in other words, complexity.[1]

Spatial design education has always dealt with problems related to real-life issues. But have the processes and products involved in spatial design education led to the creation of new knowledge or the emergence of complexities? This paper contends that hardly any such new knowledge or complexities have emerged.

Spatial designers today face far more complex spatial problems than in the past because they must also take additional factors into consideration, among them the public, social agents, civil rights agendas, urbanization, local and other regulations and new forms of knowledge and systems. The role of academia in general, and of spatial design education in particular, is to lead this change by becoming engaged in these processes in as many ways as possible. The contention of this paper is that the initiation of new approaches to education through the interrelation of transdisciplinarity and Problem-Based Learning (PBL) will lead to the emergence of new platforms and systems that synergize practice and academia. Transdisciplinarity as it relates to spatial design has been discussed recently, but not as a multiple and complex system nor in terms of the synergy between academia and practice. The new educational approaches proposed in this paper should have an impact upon both processes and products, change the perception of all agents[2] involved and ultimately transform education and practice in the field of spatial design.

In recent years both practice and academia have responded to the challenges of complex spatial problems. In *The Production of Space* Lefebvre argues that space is a social product or a complex social construction. This Marxist-humanist view affects spatial practices and changes perceptions of spatial theory and practice (Lefebvre 1974). This argument points to a shift in the research perspective, from space as a product to the processes of its production, that is, the multiplicity of spaces that are

[1]In this paper, the concept of complexity refers to that of De Landa and is in line with Deleuze and Guattari. De Landa challenges the current paradigm of social analyses to posit social entities as complexity. Complexity is emergent out of assemblages. Assemblages appear to function as a whole, but are actually coherent bits of a system whose components can be "pulled" out of one system, "plugged" into another and still work.

[2]The actor-network theory (ANT) is an approach to society that is concerned with the mechanism of power. According to this approach, society, organizations, agents and machines are all effects generated in patterned networks of (not simply human) materials. The actor-network approach is a theory of agency, a theory of knowledge and a theory of machines. More importantly, it indicates that we should explore social effects, whatever their material form, if we want to answer the "how" question about structure, power and organization. (Law 1992).

socially produced. Thus, emphasis is placed on the contradictory, conflictual and ultimately political character of the processes of production of space. The production of space belongs to a wide circle of agents, including designers, artists, social scientists and members of various communities, as well as politicians, builders, entrepreneurs, economists, policy-makers and lawyers.

These challenges generate the need to deal with complexities that "take into account the diversity of life-world perceptions of the problems" (Pohl and Hadorn 2008, 112). Therefore, complex problems can no longer be tackled through a single discipline or even multiple disciplines. This paper considers the use of transdisciplinary research (TR) methodologies as a key component in tackling complex problems and uses these methodologies as a platform for design actions. Bringing in new methodologies will require agents to rethink the relationship between process and product, leading to collaboration among multiple knowledge sources and a wider circle of agents. Previously, products and solutions were assessed according to their successful implementation. The assessment did not include questions of processes, time factors, representational tools and cultural factors relating to the various agents involved. As a result, the implementation did not take into consideration many of the factors involved, so that both product and process remained in a state of reduction and exclusion.

This paper demonstrates how new spatial design educational platforms are generated through the interrelation of TR and PBL methodologies. This model, in which the two types of methodologies are assimilated via a design participatory project, is the basis for creating new approaches to teaching and learning spatial design. Today, the boundary between practice and academia has dissolved into models of synergetic processes, allowing all agents to collaborate toward changing human well-being.

2 Transdisciplinarity and Complexity

Transdisciplinarity, as opposed to multidisciplinarity and interdisciplinarity, concerns what simultaneously exists between disciplines, crosses different disciplines and transcends all disciplines. The goal of transdisciplinarity is to understand the imperative of the unity of knowledge (Nicolescu 2005). Pohl and Hadorn claim that "the transdisciplinary challenge with complexity of problems is that of interrelating the broad range of factors to come up with integrated understanding of the problem and integrated suggestions for dealing with the problem" (Pohl and Hadorn 2008, 114). The complexity of any problem driven by current reality can only be fully understood by breaking down the subdivisions into disciplines that make it "difficult to consider an object of study as being indivisible and pertaining to only one discipline" and realizing the need to search for coherent knowledge that is not limited to a single or to multiple disciplines.

Transdisciplinary research (TR) methodology proposes four requirements for identifying, structuring and analyzing complex problems: "Grasp the complexity of problems; take into account the diversity of scientific and real-world perception of problems; link abstract and case specific knowledge; and develop knowledge and practice that promote what is perceived" (Pohl and Hadorn 2008, 112). These four requirements provide a methodological framework for bringing actors from real life into the research process and allow for collaboration and integration. The main emphasis is on research projects and the creation of new knowledge. Nevertheless, this framework remains an overall research structure comprising sequential segments that are not concerned with real projects.

As stated, in dealing with complexity, different agents from various fields of knowledge, public agencies and others must be included. Tress, Tress and Fry (2005) highlight the significance of non-academic participants in the process of connecting between academia and practice, stating that transdisciplinarity retains the same high level of scientific collaboration in academic settings as does interdisciplinarity, while also including non-academic participants in the process (Tress et al. 2005).

In this paper transdisciplinarity is examined through the theoretical framework of Pohl and Hadorn (2008), and the significance of non-academic participants is considered throughout the process (Tress et al. 2005). This combined procedure allows transdisciplinarity to be used as an applicable tool for rethinking educational approaches.

The application of transdisciplinarity approaches to the field of spatial design requires additional project-determined methodologies. Spatial design education today considers projects as comprising short or long processes and an end product. To address issues of complexity, a new agenda is needed in the form of Project-Based Learning (PBL) methodology that refers to projects beyond process and product.

3 Real-Life Project: A Case Study

In addressing real-life problems and issues of complexity in higher education, the main question is how to develop innovative integrative educational approaches to overcome the boundaries between disciplines. Steiner and Laws (2006) compared two leading universities, Harvard University (Cambridge, Massachusetts, USA) and Eidgenössische Technische Hochschule (ETH) (Zürich, Switzerland), showing how transdisciplinary methodologies changed students' perception in these institutions. In this comparison, the Harvard case study approach is referred to as the written case study approach and the ETH case study approach as the transdisciplinary case study approach.

In analyzing how both universities related to complex problems and issues and to the development of new learning formats, Steiner and Laws claimed that Harvard did not challenge current existing methods but rather focused on written cases in the form of classroom discussions. ETH, as opposed to Harvard, challenged the case study learning format by addressing related real-life problems and developing scenarios that went beyond analysis of the written case.

The ETH case study approach for creating a transdisciplinary setting included the following components: "field related knowledge, the capability to design and understand a complex system, the social competence needed to actively participate in a group together with stakeholders, the capacity to responsibly choose and apply the appropriate problem solving methods" (Steiner and Laws, p. 327). The cases focused on urban and regional problems and also on corporate sustainable development, giving "students the opportunity to attain competence in applied research in transdisciplinary setting by focusing on [a] combination of research, learning and application" (Steiner and Laws, p. 333).

This comparison of two universities is highly important in considering the issue of interdisciplinarity in higher education and in understanding its role in the interrelation between practice and academia. ETH exhibits stronger ties to all aspects of reality than does Harvard, yet both these educational formats still remain within the realm of defining processes and products as separate entities. Neither poses the challenge of creating a system in which the project becomes the central integration platform.

It appears that ETH did not fully consider transdisciplinarity when addressing issues of complexity because the projects themselves were not the main focal point of all actions and methodologies and because there was a clear separation between process and product. To bring the actual project to center stage, a new agenda is required in the form of Project Based Learning (PBL) methodologies that examine projects as more than process and product. This paper contends that only by implementing transdisciplinarity in real projects and adding new methodologies such as PBL can new educational approaches and methodologies be developed.

4 Problem-Based Learning (PBL) Methodology

PBL was researched extensively during the 1990s (Moursund 1999; Krjcik et al. 1994). In reviewing PBL, the important study by Thomas (2000) raises three significant issues. The first is assessment in order to determine what PBL is and what it is not. The second, effectiveness, questions the correct way to create efficiency models. The third addresses the issue of territory and considers whether interfaced models should be included within the PBL methodology. These three issues are critical for understanding the domain and discourse of PBL.

Thomas (2000) identifies five criteria for PBL. Some of these are essential but very basic, while others open up new possibilities and conditions but have yet to define a different kind of learning. These criteria include the following: "PBL projects are central, not peripheral to the curriculum; PBL projects are focused on questions or problems that 'drive' students to encounter the central concepts and principles of the discipline; projects involve students in constructive investigation; projects are student-driven to some significant degree; projects are realistic, not school-like" (pp. 3–4).

Ayas and Zeniuk (2001) address the issue of PBL from another perspective in the domain. Rather than considering PBL as an academically oriented approach, they see it as emerging from the practice of PBL in industrial R&D projects in companies such as Ford Motors and Fokker. By definition, these are highly complex problems and require a new learning infrastructure and a new means of implementation. The complexity of real-life projects is clearly an area that academia is only now beginning to think about in terms of opportunity rather than limitations. Ayas and Zeniuk put forward six distinguishing features of PBL for theory and practice: (1) a sense of purpose and clarity in both long- and short-term objectives; (2) psychological safety and a commitment to telling the truth as part of the project environment; (3) a learning infrastructure and a balance between emerging and formal structures; (4) communities of practice that cross project boundaries; (5) leaders who set the tone for learning and model the reflective behavior; (6) systemic and collective reflection so that problems and mistakes become opportunities for learning" (Ayas and Zeniuk 2001, 64–65).

The criterion of a sense of purpose and urgency is more easily achieved in the business world because a business must survive or else face closure. Urgency in academia can be achieved through real-life projects that take all involved stakeholders into consideration. The notion that design projects embody characteristics that grant them a feeling of authenticity was included as one of Thomas's (2000) PBL criteria. Yet this notion lacks the urgency that is so vital to complex real-life problems and to the ensuing design projects. Paraphrasing Ayas and Zeniuk, this paper proposes that short-term urgency/long-term learning be considered as one of the criteria of PBL. The criterion of urgency should drive the agents involved in the project toward creative and radical thinking. Another PBL criterion is psychological safety, which eliminates the fear of failure. Thomas claims that "projects are student-driven to some significant degree" (Thomas 2000, 4). In contrast, this paper refers to Ayes and Zeniuk and proposes eliminating the hierarchy and rethinking involved in evaluation and assessment so as to encourage natural leadership as opposed to management roles. Ayes and Zeniuk state that leadership "reflects the emergent structure and the evolving culture … The leader as reflective practitioner sets the tone for learning" (Ayes and Zeniuk 2001, 72). Academic organizations that promote degree hierarchy and teachers that insist on superior models encourage practically no organic leadership. Examples of academic organizations that do promote organic leadership can be found in some design labs, such as the MIT Media Lab.[3]

While Thomas's (2000) academic model refers to a single project framework, Ayas and Zeniuk's (2001) model suggests that projects comprise communities of practice that allow all individuals to belong to multiple communities. In academia,

[3]Torjman, L.: Labs: Designing the Future. MaRS Solution Labs Report (2012). The labs in general, and the media lab in particular, offer a neutral space dedicated to problem-solving in a highly experimental environment. Projects that are initiated by the needs of industries result in prototyping, allowing a group of students, instructors and users to learn by doing rather than by thinking. Each of the projects employs a user-centric lens, making the end-user into a critical participant throughout the process. The labs focus on diversity of perspectives and skill sets, and on team process, thus representing a convergence of design, ethnography and business to support both theoretical and real-world applications. In a lab, the whole (that is, the solution) is greater than the sum of its parts (the input of individual participants). Proprietary ownership is minimized in favor of objectivity and a commitment to a shared vision.

this model includes different kinds of courses that interrelate and form a network of knowledge, systems and human resources.

Reflexive acts are inherent to the culture of design teaching and learning, and are manifested in all courses in the curricula. It is the contention of this paper that reflexive acts should break the boundaries of the design curricula by transforming spatial design education into a collaborative and participatory design action that includes all agents involved, among them students, faculty members and various communities. This action is referred to as Participatory Reflexivity.

This approach to spatial education adopted here involves assimilating transdisciplinary methodology and PBL methodologies. The following section demonstrates the above proposal using an example of a unique spatial design project called SpeeDesign.

5 Project-Based Design Case Study—The Designers Clinic and SpeeDesign

The Designers Clinic is a unique platform positioned between design practice and academia whose objective is to merge the field of spatial design and the world of social engagement. The platform uses a bottom-up design approach based on community participation in design processes and products. The Designers Clinic operates as a flexible axis running through various academic degree formats and design practice platforms. The clinic attracts multiple stakeholders from such diverse perspectives as academic agents, social and governmental agencies and non-government organizations (NGOs). These agents are the clinic's clients and at the same time the participants in numerous design and planning projects.

The Designers Clinic is a unit within the academic system that initiates projects through educational mechanisms such as studio classes, workshops, internships and labs. The various projects promote the development and innovation of social design and entrepreneurial knowledge by working with academic faculty and experts from various fields of expertise. Promoting project excellence requires the collaboration of experts in design management and entrepreneurship, as well as innovation processes and expertise in community participation methodologies. In addition, the clinic focuses on establishing ties with the community and on identifying changing values and needs contributing to the well-being of various communities.

SpeeDesign is a unique entrepreneurial project run through the Designers Clinic. The first collaborative and participatory SpeeDesign project was held at the Peila Cultural and Art Community Centre (NGO) located in the less privileged old Arab quarter of the Tel Aviv-Jaffa municipality. The end product of the project was an event at which 60 top interior designers, architects and engineers offered design and planning services to members of the local community (Figs. 3 and 4). At the specified time, the designers and architects sat at tables with empty chairs, waiting for something to happen. This was an awkward moment because nobody knew whether community members would be active participants in the process or would stay away, choosing not to participate.

An hour later, the space was packed with people, including families from Jaffa, single people from Tel Aviv and representatives of small businesses in the area. Some brought scraps of paper on which they had scribbled the dimensions of their apartments, others came with architectural plans that highlighted the furniture within the space and still others brought their iPods (Figs. 1 and 2). Some brought photographs, either hard copies or stored on their mobile phones. The professionals happily took every piece of information they were given and used it productively. The community

Fig. 1 Speed Design—Architectural information prepared in advance by a resident

Fig. 2 Speed Design—Designers and member of the community discuss a plan

Fig. 3 Billboard advertising the event, displayed all over the city

Fig. 4 Speed encounters between designers and the community

participants were ecstatic and moved from one consultant to another, receiving design solutions and plans of their living and working spaces. Architects with different areas of expertise referred people to their colleagues. Food and beverages were served by local community restaurants and stores. The event felt like a big festive happening, and many people called their friends to come and participate. It should be noted that an entire year of research, planning, production and collaboration was needed to create the infrastructure that made an event of this nature possible.

The SpeeDesign project has all the characteristics of TR methodologies. The process of grasping the complexity of people's right to housing is manifested in the need to combine a wide range of housing issues with differing perceptions, agendas and motives. Development and innovation in the project involved various methodologies for researching multiculturalism, housing typologies and community engagement. Identifying the above issues required taking into account everyday practical constraints to solve urgent residential problems. The process and product development led to the emergence of a network comprising faculty members, professionals, social activists and citizens, representatives of local authorities and others.

An examination of the criteria of PBL indicated that SpeeDesign is also a perfect PBL case study. One of the PBL criteria met by the project is its unique position, in which processes and solutions are simultaneously both short-term and long-term. Engaging with communities usually takes a long period of time and considerable effort. By means of rapid design actions, SpeeDesign made it possible to empower people and bring about significant changes in their well-being and in their living environment. Long-term refers to continuous efforts toward developing design strategies, business plans, communication strategies and joint cooperation with all communities involved in order to generate long-term housing solutions.

The SpeeDesign model for interrelating TR methodologies and PBL also had an impact on the spatial design curriculum and became an integrative model within the undergraduate and graduate studies programs in the form of real-life projects. Projects such as SpeeDesign became the focal point of all academic courses, among them business courses teaching all aspects of the business plan, from marketing, sales and branding to public relations. Other courses addressed questions regarding community identification, strategies for reaching out to communities and engagement in participatory models.

A reflexive process was initiated to develop a transdisciplinary and PBL model of teaching and learning. This process required the ongoing participation of all agents involved. After the event, the participating designers were asked to complete questionnaires that included the following four questions:

1. In helping the community, to what extent did you grasp the problem of the "right to housing"?
2. To what extent did you feel that you could have an impact on the housing conditions of all participants in order to improve their well-being?
3. What kinds of new design tools are offered by the SpeeDesign format?
4. To what extent did your participation change the way you perceive your profession?

All the professional designers and experts stated that focusing on the problem of housing through a different kind of real-life project like SpeeDesign required them to rethink their approach to the issue of complexity. They also stated they were surprised that they were able to help every one of the participants, regardless of the design problem and regardless of the information brought to the table.

In response to the fourth question, one designer said: "The entire encounter was a promotional collaborative act for the professional field of spatial design. To my surprise, the collaboration between different experts could actually generate positive change for the well-being of all kinds of residents." Another architect wrote: "The possibility of such a quick and accessible interaction is not only productive for the citizens but is also a highly liberating experience for the giver (the professional)."

One of the findings of the reflexive analysis showed that among the professionals involved, educators grasped the opportunity to use PBL for cultivating awareness of social and ethical dilemmas among future designers. This project signified the importance of dissolving the boundaries between academia and practice by changing the perception of these educators.

One of PBL criteria is the creation of leadership that is organic and not hierarchic in order to change decision-making procedures and relations among all members, from academic faculty members and students to members of the community and professionals. One example of how the existing hierarchy and decision-making processes were broken down is that each person from the community was given multiple ideas and designs and communicated with different professionals, rather than being given a single design solution for an urgent housing problem. In design and architecture practice, multiple ideas and second opinions are usually considered unethical and are generally prohibited. The process challenged this norm and led to the emergence of an ethical principle for both process and product to create multiplicity in the identities of interiors.

In assimilating PBL and TR methodologies and criteria, a new teaching and learning culture emerges, one that promotes reflection, debate and a new way of thinking and acquiring knowledge.

6 Discussion

The development of new approaches to teaching and learning by means of real-life complex problems will ultimately change the shape of spatial design education. This paper proposes a foundation for new spatial design methodologies based on the synergy between transdisciplinary methodologies and PBL. The paper demonstrates that transdisciplinary methodologies remain theoretical in that they do not use the project as a mechanism for integrating all aspects and agents involved. Hence they are insufficient to deal with and implement complex, real-life problems. Only by introducing risk-taking PBL methodologies, generating organic leadership and promoting short- and long-term learning with transdisciplinary performance components can real-life design projects be used to solve problems and empower all involved stakeholders. As such, the SpeeDesign platform, in the form of a real-life project, is an example of how a spatial design engagement project becomes the focal platform for transdisciplinary approaches. For collaborative projects such as SpeeDesign to transform spatial design education and remove the boundary between academia and practice, a reflexive process and debate must be established and implemented by all spatial design disciplines, including architecture, interior architecture, design and urban design practices, and by all practitioners as well, among them artists, social scientists and communities, politicians, builders, entrepreneurs, economists, policy-makers and lawyers.

The real-life project can become a force for innovation and can change the perceptions of all involved agents, as well as the overall perceptions, content methodologies and outcomes of the discipline as a whole.

References

Ayas, K., Zeniuk, N.: Project-based Learning: Building Communities of Reflective Practitioners. Management Learning. 32 (1), 61–76 (2001)

De Landa, M.: A New Philosophy of Society: Assemblage theory and social complexity, London & New York: Continuum (2006)

Krajcik, J. S., Blumenfeld, P. C., Marx, R. W., & Soloway, E.: A Collaborative Model for Helping Middle-Grade Science Teachers Learn Project-Based Instruction. The Elementary School Journal. 94, 483–497 (1994)

Latour, B.: "A Cautious Prometheus? A Few Steps toward a Philosophy of Design." In Networks of Design, edited by Jonathan Glynne, Fiona Hackney and Viv Minton, 2–10. Boca Raton, FL: BrownWalker Press. (2008)

Latour, B: Reassembling the social—An Introduction to Actor-Network-Theory. Oxford University Press (2005)

Latour, B. 2009. "Spheres and networks: two ways to reinterpret globalization." Harvard Design Magazine 30: 138–144

Law, J.: Notes on The Theory of the Actor-Network: Ordering, Strategy, and Heterogeneity. Systems Practice. 5 (4) 379–393 (1992)

Lefebvre, H.: The Production of Space. Blackwell Publishing (1974)

Moursund, D.: Project-based Learning Using Information Technology. Eugene, OR: International Society for Technology in Education. (1999)

Nicolescu, B.: Transdisciplinarity—Past, Present and Future. CETRANS—Centro de Educacao Transdisciplinar (2005)

Pohl, C., Hirsch Hadorn, G.: Principles for Designing Transdisciplinary Research. oekom Verlag, Munchen (2007)

Steiner, G., Laws, D.: How appropriate are two established concept from higher education for solving complex real world problems?: A comparison of the Harvard and the ETH case study approach. International Journal of Sustainability in Higher Education. 7 (3), 322–340

Thomas, J.: A Review of Research on Project-Based Learning. https://www.bie.org/index.php/site/RE/pbl_research/29 (2000)

Torjman, L.: Labs: Designing the Future. MaRS Solution Labs Report (2012)

Tress, G., Tress, B. & Fry, G.: Clarifying Integrative Research Concepts in Landscape Ecology. Landscape Ecology. 20 (4), 479–493 (2005)

Global Designers and Local Community: Bridging the Gap Through Information and Communication Technologies

Yael Valerie Perez, Angela James, Erica Carson, and Alice Merner Agogino

Abstract This research uses an international design competition as a research platform to evaluate the effectiveness of Information and Communication Technologies (ICTs) in transdisciplinary design. Specifically, it focuses on transcending cultural gaps between professional designers and the local community of users. While ICT has been speeding global processes and facilitating design across professional disciplines, its capacity to support collaboration and communication among professional and non-professional stakeholders is questioned. Through an international design competition, with a Native American Nation in Northern California, the design process is evaluated and the solutions created by distant designers, who had only an ICT-mediated experience, is compared with those of local designers who had an additional in-person experience. Furthermore, the designers' level of familiarity with the place is evaluated before and after the ICT interventions are introduced. The research reveals that a place-specific set of ICTs, created by the local community to represent their conceptions of place, can transcend not only the geographic distance, but also disciplinary and cultural gaps. The conclusions highlight the importance of a transdisciplinary design processe, that involves local, non-professional experts in the creation of the solution.

Keywords Design methods · Co-design · Social factors in design · Community-based design · Community-based research · ICT · ICT4D · Design competition

Y. V. Perez (✉)
Blum Center for Developing Economies, UC Berkeley, Berkeley, USA
e-mail: pyael@berkeley.edu

A. James · E. Carson
Tribal and Historic Preservation Office, Pinoleville Pomo Nation, Ukiah, CA, USA

A. M. Agogino
Mechanical Engineering Department, UC Berkeley, Berkeley, USA

L. Blessing et al. (eds.), *The Future of Transdisciplinary Design*,
https://doi.org/10.1007/978-3-319-06382-9_18

1 Background

1.1 Introduction

This community-based research evaluates the use of Information and Communication Technologies (ICTs) as supporting tools for transdisciplinary design. The goal of this evaluation is to identify tools to transcend wide cultural and geographical gaps between design professionals and the local community of users, as nonprofessional experts on the place. The study measures the effectiveness of ICTs, in communicating information about the place, and determines their capcity to allow architects, who cannot access the place, equipotential opportunities to produce place-appropriate designs as those who were able to visit the place and meet with users face-to-face. The place-appropriateness of the designs are evaluated both by professional designers and by community representatives.

Situated in place-making and environmental psychology fields of study, the research relies on Canter's definition of place as the overlap between physical attributes, activities, and conceptions (Canter 1977). To learn conceptions of place—the subjective understandings of local users—interaction between designers and local users should occur. Therefore, using Blessing's design model inclusive of strategies, stages and activities (Blessing 1995), transdisciplinary design, as a *strategy* for place-appropriate results, requires local users to work with professional designers throughout the design *stages* and across the variety of design *activities*.

While the information superhighway and general globalization processes reduce cultural differences between places and people, they also allow more designers to design in places that are geographically and culturally distant from their home base. As a consequence, small and rural communities are no longer restricted by the limited number of professional designers they can find in their area. Therefore, the challenge of understanding local characteristics and communicating these within teams of professionals and nonprofessionals at a distance is critical. As such, transdisciplinary design has the power to balance global forces with unique, local attributes and needs.

The international design competition that was used as a research platform was organized by CARES,[1] a team of faculty and students from the department of architecture and Mechanical Engineering at University of California at Berkeley, and the Pinoleville Pomo Nation (PPN), a small Native American (NA) Nation located in Mendocino County in Northern California U.S.A. The conceptual design competition—ParticiPlace2012[2]—was opened to practicing designers and students from all over the world. The competition challenged designers to compose a conceptual design for a Living Culture Center for the PPN. This vision for the building was to serve as a space to practice, preserve and present the culture of the tribe. The design brief for the competition was created by the CARES team together with tribal citizens over a series

[1]CARES: Community Assessment of Renewable Energy and Sustainability. For more details about CARES, see planetcares.org.

[2]For more information 2012.participlace.org.

of workshops and meetings. The competition attracted some local designers who were able to attend site-visits organized by the team, and other designers who were not able to attend the site visit and had to rely solely on ICT-mediated experiences to learn about the place and the people.

1.2 Top-Down and Bottom-Up Approaches for Bridging Cultural Gaps with US Native American Nation

Emphasis on the user experience in the environmental design literature is strongly represented by the social design movement in architecture. Ethnographic methods, such as observations and interviews are often used in different design fields such as engineering, industrial design, and architecture. Yet many of them leave the user passive. More active methods—such as design charrettes, a collaboration method common in the architectural realm, or co-creation developed in engineering—are often time-consuming and their long-term financial benefits are hard to justify. To encourage architects to add local users into the transdisciplinary design process, a strategy combining both political and academic approaches is required. In the political realm changes are required in building policies to empower local communities of users and to provide them with freedom to make their own design decisions at the local level. In 1996 the U.S. department of Housing and Urban Development (HUD) passed the NA Housing Assistance and Self-Determination Act (NAHASDA), which allows NA Nations to use federal funds to self-compose their housing solutions as opposed to being limited to pre-designed HUD houses.[3] This top-down change was the starting point of our co-design project with the Pinoleville Pomo Nation (PPN), which opened up the opportunity to identify the tools to support transdisciplinary design with emphasis on local experts and users. Technology by itself, we acknowledge, may not be enough for transdisciplinary design that includes local users. Design that allows local users to control decisions and project budgets, on the other hand, has the potential to become transdisciplinary in a way that can be facilitated with effective technology. The PPN's search for ways to use federal funds to create housing that supports their unique culture and desire for self-sustainability led them to CARES at UC Berkeley.

Due to their unique culture and lifestyle within U.S. society, the PPN is part of a place that is unfamiliar to many designers, both within and outside the U.S. Being experts in their unique place, yet with no professional design expertise, PPN representatives constitute a range of disciplines concerning their cultural norms, local environment and user needs, therefore making them a comprehensive part of transdisciplinary design. Technology to support this type of transdisciplinary design has to

[3]The two programs authorized for Indian tribes under NAHASDA are the Indian Housing Block Grant (IHBG), which is a formula-based grant program, and Title VI Loan Guarantee, which provides financing guarantees to Indian tribes for private market loans to develop affordable housing. Regulations are published at 24 CFR Part 1000 (HUD 2012).

transcend geographical, cultural and disciplinary gaps. In-addition, the very limited natural and financial resources of the PPN and the lack of professional services within their own culture, required us to focus this investigation on the extreme condition of a marginalized place with the underlying assumption that technology that can transcend these gaps will likely to be effective in narrower gaps as well. These extreme design conditions and gaps often result in a top-down approach overlooking different place-specific elements, resulting in culturally-inappropriate designs as in the examples provided by Rapoport (1990). Technology to bridge the designer—with his foreign, professional, top-down approach—and the community—located at the place, looking bottom-up—should support communication and sharing of information.

1.3 Literature Review—Technology to Support Comprehending Conceptions

Most literature investigating the use of ICT for involving nonprofessionals in the design process focus on communicating physical-attributes of place, particularly the designed place. These are often missing Canter's activities and conceptions (Canter 1977) for a complete place representation. Both these additional elements are directly connected to the user experience and transdisciplinary design that involves local expert users, in the early design stages, provides a good strategy for the designer to comprehend these richer aspects of place. The inclusion of local, nonprofessional people in collecting and representing place-related-data is already found in the field of Participatory GIS (Van Wart et al. 2010; Dunn 2007; Abbot et al. 1998).

Three-dimensional digital environments have been found effective in research looking into way-finding (Bhatt et al. 2011) and have been extensively used in the fields of architectural and urban history (Michon an El Antably 2013, El Antably 2010; AlSayyad 1999, AlSayyad et al. 1996). In recent years, the development of multiusers virtual environments (MUVEs) has allowed limited representation of people through embodiment (Chen and Kalay 2008) as part of place and has been tested and proven effective in journalism (Kalay and Grabowicz 2006) and historic preservation (Michon et al. 2008). MUVEs rely on gaming technology and producing a place-specific MUVE is time consuming and requires some professional skill. Online MUVEs, such as Second life, have been effectively used in design education (Hong et al. 2013) mostly for virtual environments and not as a representation of an existing environment.

Based on these experiences we developed the ParticiPlace methodology, which includes the media of images and videos that can easily be created by and with non-professional local experts to represent conceptions. The users are thus able to express the place using these representations as cultural probes.

Cultural probes, as developed by Gaver, Dunne and Pacenti (1999), are tools, such as postcards and cameras, to gather subjective information from users as inspirational material to familiarize the designer with groups of users. Originally developed for art and architectural design and inspired by Situationist International, the concept has recently been adapted by designers of Human Computer Interaction (HCI) as an alternative option for knowledge production (Boehner et al. 2007). Dealing with subjective information on the user's side and inspiration on the designer's side make these tools hard to evaluate (Gaver et al. 2004). Our case study uses digital cultural probes as representations of conceptions of place. The competition setting allows us to include an empirical evaluation of the method, using a few probes created with a community of users, to inspire a variety of designers working on the same project. We then evaluate the subjective and objective influences of the probes on the design process and compare the responses of those who were able to include a face-to-place interaction to those restricted to the digital probes.

2 Approach

2.1 Design Competition as a Design Experiment

This research uses ParticiPlace, a community-based, international design competition, as the main platform to explore the effectiveness of ICT in producing place-appropriate design. Other examples of research using design competitions include Wooten and Ulrich (2013), focusing on innovation, and Valkenburg and Dorst (1998), focusing on collaboration within design teams.

Design competitions present a real life design scenario in which different designers work on the same project at the same time and thus can be defined as in vivo experiments. Advantages of using a community-based, international design competition as a research platform include:

- Real life conditions, with prizes and reputation at stake, encourage designers to produce the best designs they can.
- Competitions are a common scenario in global architectural practice, in general, and in developing regions, in particular. Many international competitions are used in developing regions to elicit a variety of contemporary international design ideas.
- They provide access to actual users/clients in relation to an actual project and designers, crucial components for evaluating ICT to support transdisciplinary design.
- Some evaluation of the designs is already embedded in the competition process.

Alternative research methods such as design experiments (e.g., Al-Sayed et al. 2010) would be restricted to design in the "lab" and would therefore use a much smaller time frame. Experiments based on design studio-classes (e.g., Bakergem in McCullough, Mitchell and Purcell 1990) often have longer timeframes yet are restricted to design students and would rarely include users or clients, an important component of conceptions. Shah et al. (2003) offer a design research method to overcome the science-engineering dichotomy and align engineering, experimental research methods with general and simplified lab experiments used in the sciences. They suggest decomposing each design method tested into key-components. The overall effectiveness of a method can then be predicted by experimentally studying the effectiveness of its components and its mutual interactions. Although such approaches can be used in constrained applications, real life transdisciplinary design has characteristics of complex systems with complex interconnections between its components, thus ICT successful in an in vitro experiment may fail in vivo. Therefore a methodology to evaluate the overall design strategy in "living" design conditions is needed.

Research focused on real-life conditions and looking at practiced design methods in industry often uses qualitative methods such as "building stories" as an investigation method (Martin et al. 2005). The narrative technique allows capturing aspects of practice in different levels (overall project, team level and individual). With the goal to retain the realistic context of design, the research relies on the competition as a case-study which includes a small sample of participants but multiple sources of evidence, both qualitative and quantitative. These evidences capture both the nomothetic general properties influencing the design process (through questionnaires and blind evaluations) as well as the idiographic, specific characteristics recognized by individual designers (through questionnaires and interviews).

2.2 The Competition Process

The competition announcement was sent to architecture schools around the globe, to international architecture and architectural-competition journals and websites, as well as to a variety of architecture firms in California and elsewhere. In addition, the call for proposal was spread through personal contacts. Transdisciplinary teams were encouraged by including, in the competition brief, challenges pertaining to a variety of disciplines, by providing separate prizes for sustainable engineering innovation and social and cultural integrity, and by including a multidisciplinary team of experts in the jury.

Thirty-eight groups registered to the competition, of which 17 submitted a design: seven from California, three from other U.S. states, and the rest from outside the U.S. The jury was comprised of three practitioners in architecture, engineering and environmental science and two PPN representatives. The jury members evaluated submissions through evaluation forms tailored for this competition, which were aggregated to calculate the winners.

Communication with registered participants of the competition was managed through a Google group and participants were encouraged to send questions throughout the process. The 17 design submissions represented 56 individual designers, of which 12 representatives from six groups were able to attend the site visit. The vast majority of the designers were architects; a few had previous professional experience in other disciplines: one team had a landscape architect, another included an engineer.

2.3 *The Treatment: Technology Provided to Support Transdisciplinary Design*

To support transdisciplinary design with a focus on the local expert users, a separate process was conducted with the local community of users to communicate conceptions of place. The goal was to engage the community in different ways that would support communication of their conceptions of place. We introduced a variety of available technologies—such as photo editing, video editing, blog writing, and social networks—through a workshop we conducted with about 15 PPN representatives—young children, youths, and adults. We created videos of PPN leaders walking the land and presenting the important qualities. The PPN representatives described the problems of performing cultural activities in current available spaces. In addition, the PPN representatives created a video in which they presented a variety of cultural aspects relevant to the living culture center. These videos, as well as several sets of photos of cultural activities, were made available for participants during the design process. They were posted on Youtube and Flickr and were open to comments. Additionally, a dedicated Facebook page and Twitter account were set up for the competition and made available for competitors and local PPN citizens to join and follow. The photos and videos were posted through these social media as well. Designers were asked to evaluate their familiarity with each element of the place and attribute the basis for this familiarity, both before the use of these additional technologies and after using (or not using) them for one month. In addition local representatives of users evaluated the submitted designs and their appropriateness to the place.

2.4 *Data Collected*

Throughout the competition the following methods were used to collect data:

- Initial registration form (37 teams out of which 17 submitted designs).
- First questionnaire: Self-evaluation of familiarity with the place and its sources *before* additional media (11 respondents out of 37 teams—30% response rate).
- Second questionnaire: Self-evaluation of familiarity with the place and its sources *after* additional media (13 respondents—35% response rate, of which 8 also answered the first questionnaire. 7 teams responded from the 17 who submitted designs—41% response rate).
- Communication through Google groups (13 questions).
- Interviews with participants (6).
- Interview with Jury (2).
- Online hits and posts on Flickr, YouTube and Facebook.
- 17 × 5 Jury evaluations for each submitted design (5 jury members—2 non-professional, local experts and 3 non-local design professionals) included 17

questions pertaining to the design requirements and its appropriateness to each element of the place.

3 Findings

3.1 Who Are the Participants and What Do They Know About the Place and the Users?

None of the participants had been to the PPN's reservation before the competition. Yet, 56% had been on other NA reservations. Other participants came into the competition with very little experience of the place and its people: 18% had never been to California, nor had met NA people of California.

Participants were asked to rate their familiarity with different elements of the place on a scale of 1 to 7 (very low familiarity to very high familiarity). Then they were asked to assess their place familiarity by selecting from a list of options that will be discussed in the following section. The subjective feeling of initial familiarity with the place is presented in Fig. 1, in comparison to the objective experiences of the designers (been or never been to a NA reservation in California). As can be expected, those who had never been to a NA reservation in California graded their familiarity lower than those who had. It is interesting to see that the difference between those who had the experience and those who didn't was greater in the "familiarity with people" than in the "familiarity with physical properties" of the place or the activities. Moreover, designers who had been to a NA reservation in California (mostly people who lived in California) felt that they knew the people more than the activities; yet, those who hadn't visited (or lived in California) felt that they knew the activities better than the people. Those who have been to a NA reservation felt their familiarity with activities was their weakest point. This shows the difficulties designers face in learning about the people for whom they are designing and learning their conceptions of place when there are large cultural gaps.

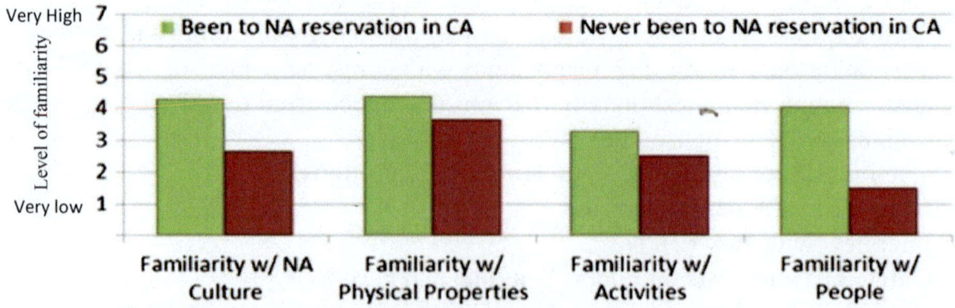

Fig. 1 Designer's self-evaluations of level familiarity with different elements of the place

3.2 *What Did the Participants Ask About the Place?*

The design brief for the competition included information about the physical attributes (such as maps and weather information), activities (such as a bubble diagram of the required activities), and conceptions (such as a short history of the land as told by the PPN representative, moods and characters identified by PPN citizens for each required space). In addition, the design guidelines as chosen by the PPN representatives were the Fully Integrated Thinking ™ (FIT) framework developed by the Biomimicry 3.8 and HOK, which includes 15 lenses touching all three elements of place. Beyond the information provided, a total of seven e-mails with 13 questions were sent from six different participants and 33 questions were posted through the registration form. Sixty-five percent of the questions related to physical attributes such as asking for site photos, additional climate data, geologic data etc. Nine percent related to activities and included requests for images of actual tribal members, their representations of place, as well as asking for more details about square footage of the different space/activities defined in the program. Finally, 11% related to conceptions and asked about cultural beliefs, or tribal stories associated with a link one of the participants found online. The rest were related to submission instructions and not to the place.

Based on discussions with participants, after the competition ended, the design guidelines were very detailed for the activities, which may explain the low number of questions asked in this realm. By providing characteristics and moods of each space, the design brief provided more information about the conceptions of place than most design projects, and only a few architects required more details about this element.

3.3 *Media*

A comparison of the self-reported evaluation of familiarity with the place by designers who didn't visit the site, before and after being introduced to the added media (one month apart), provides some insight into ICT's influence. On average the reported familiarity with the physical attributes grew by 7.14%, activities grew by 14.3% and conceptions grew by 8.6%. After submitting their designs, participants were asked to evaluate each media used on a scale from 0 (not effectual) to 4 (extremely effectual), as presented in Fig. 2. According to the designers who had an additional non-mediated interaction this experience was the most influential on the design (some influence was found when another team member visited the site within the design team). In general, most media were more influential for those who only had the ICT-mediated experience. This emphasizes the perceived importance of such alternative sources of information by designers who have no access to in-person experiences. Books, movies (non-online) and weather data were still found to be more important to those with direct experience. This may imply that these media are more accessible to local designers and that non-local designers may have language barriers or difficulties

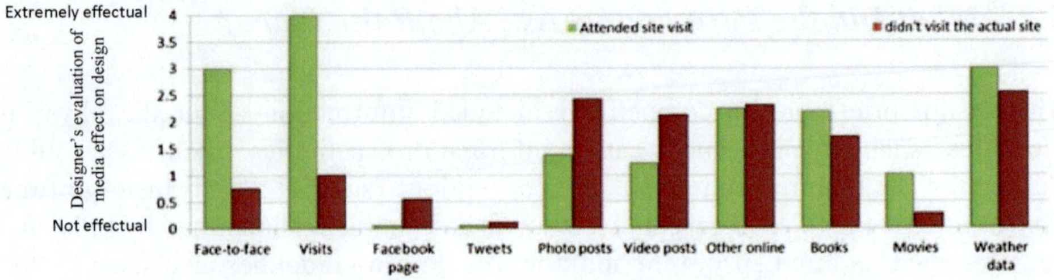

Fig. 2 Effect of different information types on the design based on designer's experience

identifying or accessing relevant information through this media. Moreover, it shows that those with limited direct access to place and people relied mostly on provided online sources.

3.4 "Meeting the Place"—A Transdisciplinary Evaluation of Design

As part of the evaluation of place, Canter suggests considering the evaluations that people make. Hence, the evaluation of the submitted designs included a transdisciplinary jury of professional (architects, engineers and biologist-designer), as well as local, representatives of the PPN. The form provided to the jury included grading a list of 17 criteria, taken from the design brief, including specific questions about the appropriateness of the design to the PPN. Subjectively, ICT successfully contributed to the familiarity with the place of those who couldn't visit the site. Additionally teams who visited the place, performed, on average, as well as those who relied solely on ICT-mediated experiences. The variations in the evaluations between the PPN representatives were high, ranging from a standard deviation of 0 (consensus) to 40 while the standard deviation of the professional team ranges from 2 to 21. The PPN jury members also had a higher standard deviation within their own evaluation compared to the professional jury members. This calls for a larger jury or evaluators, particularly from the nonprofessional members.

Besides the 14 questions in the jury's evaluation forms focusing on different qualities of the design corresponding to each of the FIT lenses and the design brief, three additional questions asked the jury to evaluate, on a 1–7 scale, the "appropriateness of the design to the physical properties of the place", "to the activities" and "to the people". Theses evaluations were compared to the self-evaluation that designers provided. Very weak correlation was found between designers' perception of familiarity with the place and the jury's average evaluation. Nevertheless, when separating the jury into the professional jury and the local experts, a different picture appears (Fig. 3): While the local experts had little correlation with the designer's self-evaluations, the professional jury is very closely correlated to the self-evaluation of the designers. This provides evidence of the disconnect between the professional

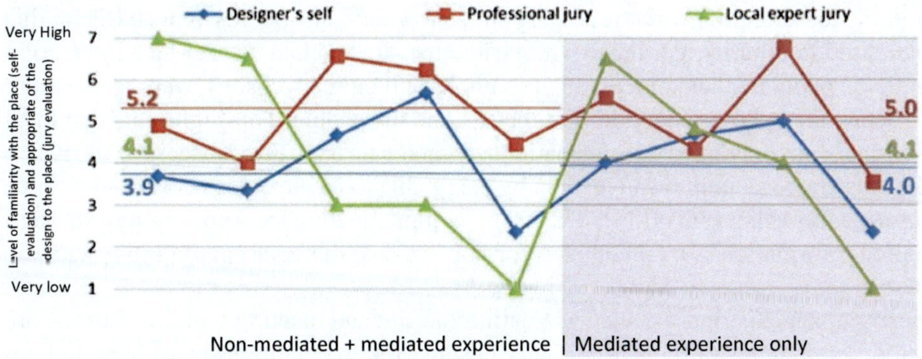

Fig. 3 "Meeting" the place- comparing evaluations of professional jury, nonprofessional jury, and designers

designers and the local experts, as well as the connection between the competing designers and the professional designers in the jury: When a participant felt familiar with the place, whether based on direct or IT-mediated interaction, they also managed to convince the professional multi-disciplinary jury that they were familiar with the place, but not necessarily the local, non-professional, community experts.

4 Conclusions

This research relies on an environmental design understanding that includes conceptions—local-user's subjective understanding of place—as one of the elements of place (Canter 1977). Conceptions would be hard to communicate without some involvement of the local users. Therefore understanding conceptions as part of place-appropriate design calls for a transdisciplinary design process that brings together designers and local expert users.

The ParticiPlace paradigm used in this case study includes information technologies, such as digital photos and videos produced by or with the community and communicated through different web-based social media such as Facebook, YouTube, and Flickr. This set of technologies aims to reflect the conceptions of the local community of users regarding the (current) place and the project (future place).

The research shows that ICT helped bridge the gap between non-local designers and place. Nevertheless, based on objective usage data and subjective reports provided by the designers through questionnaires and interviews, not all media were effective for the design. Facebook, as a social network, proved to have only a slight influence on the design. Most designers claimed that they were not regular users of Facebook or other social networks. The local community had concerns relating to privacy that limited their choice to communicate through this medium as well. Measuring subjective evaluations of familiarity with the place during the transdisciplinary design shows that familiarity grew over time. Designers who were initially least familiar with the place, pointed to videos and photos as the basis for this increased familiarity. For those who visited the site and had face-to-face interaction with the people, other sources, such as books and online websites, were found more effectual than photos and videos of the site and the people. This highlights that local designers not only have an information advantage through direct access to the place, they also have an indirect advantage by being able to locate other relevant sources of information online and offline. Therefore, communicating local conceptions through online information is a crucial step for non-local designers to get familiar with the place.

In-person experiences, such as visiting the site and meeting with local users, are important features of transdisciplinary design. Yet, in conditions of wide geographical and cultural gaps, the ParticiPlace paradigm, in which local users use ICT to represent their place and their conceptions, can help overcome the gaps and allow both local and non-local designers to become familiar with the place. This paradigm is project- and place-specific and its time-efficiency compared to face-to-face interactions was not evaluated as part of this research. It would therefore be wrong to assume that ICT should replace in-person interaction in transdisciplinary design.

An interesting correlation was found between designers' self-evaluation of knowing the place and the final-design evaluation of the professional jury while only a weak correlation was found with the final-design evaluation of the nonprofessional, local expert jury. Future research should focus on transcending place conceptions between professionals and non-professionals.

Bibliography

Abbot, Jo, Robert Chambers, Christine Dunn, Trevor Harris, Emmanuel de Merode, Gina Porter, Janet Townsend, and Daniel Weiner. 1998. "Participatory Gis Opportunity or Oxymoron." PLA Notes 33 (October): 27–33

Al-Sayed, Kinda, Ruth Conroy Dalton, and Christoph Hölscher. 2010. "Discursive Design Thinking: The Role of Explicit Knowledge in Creative Architectural Design Reasoning." AI EDAM 24 (Special Issue 02): 211–230

AlSayyad, Nezar. 1999. "Virtual Cairo: An Urban Historian's View of Computer Simulation." Leonardo 32 (2): 93–100

AlSayyad, Nezar, Ame Elliott, and Yehuda Kalay. 1996. "Narrative Models: a Database Approach to Modeling Medieval Cairo." ACADIA'96

Bhatt, Mehul, Christoph Hölscher, and Thomas F. Shipley. 2011. "Introduction and Symposium Overview." In Spatial Cognition in Architectural Design. New York

Blessing, L. T. M. 1995. "Comparison of Design Models Proposed in Prescriptive Literature". Book Section

Boehner, Kirsten, Janet Vertesi, Phoebe Sengers, and Paul Dourish. 2007. "How HCI Interprets the Probes." In Proceedings of the SIGCHI Conference on Human Factors in Computing Systems, 1077–1086. CHI '07. New York, NY, USA: ACM

Canter, David V. 1977. The Psychology of Place. Architectural Press

Chen, Xiaole, and Yehuda Kalay. 2008. "Making a Livable Place: Content Design in Virtual Environments." In New Heritage: New Media and Cultural Heritage, edited by Yehuda Kalay, Thomas Kvan, and Janice Affleck. Routledge

Dunn, Christine E. 2007. “Participatory GIS — a People’s GIS?” Progress in Human Geography 31 (5) (October 1): 616–637

El Antably, Ahmed. 2010. “The Virtual (Re)Construction of History: Some Epistemological Questions” (September 15). https://escholarship.org/uc/item/225426rr

Frischer, Bernard. 2005. “New Directions for Cultural Virtual Reality: A Global Strategy for Archiving, Serving, and Exhibiting 3d Computer Models of Cultural Heritage Sites.” In Proceedings of the Conference, Virtual Retrospect 2005 (Bordeaux), 168–175. Biarritz—France

Gaver, Bill, Tony Dunne, and Elena Pacenti. 1999. “Design: Cultural Probes.” Interactions 6 (1) (January): 21–29

Gaver, William W., Andrew Boucher, Sarah Pennington, and Brendan Walker. 2004. “Cultural Probes and the Value of Uncertainty.” Interactions 11 (5) (September)

McCullough, Malcolm, William John Mitchell, and Patrick A. Purcell. 1990. The Electronic Design Studio: Architectural Education in the Computer Era. MIT Press

Michon, Daniel, and Ahmed El Antably. 2013. “It’s Hard to Be down When You’re up: Interpreting Cultural Heritage through Alternative Media.” International Journal of Heritage Studies 19 (1): 16–40

Michon, Daniel, Kalay Yehuda, Selina Lam, Yael Perez, and SeungWan Hong. 2008. “Virtual Sambor Prei Kuk: Weaving The Tangible and Intangible Cultural Heritage.” In Virtual Sambor Prei Kuk: Weaving The Tangible and Intangible Cultural Heritage. Limassol, Cyprus: VSMM Press

Rapoport, Amos. 1990. The Meaning of the Built Environment: a Nonverbal Communication Approach. University of Arizona Press

Shah, Jami J., Steven M. Smith, Noe Vargas-Hernandez, David R. Gerkens, and Muqi Wulan. 2003. “Empirical Studies of Design Ideation: Alignment of Design Experiments with Lab Experiments.” In ASME Design Theory & Methods Conference

Valkenburg, Rianne, and Kees Dorst. 1998. “The Reflective Practice of Design Teams.” Design Studies 19 (3) (July): 249–271

Van Wart, Sarah, K. Joyce Tsai, and Tapan Parikh. 2010. “Local Ground: a Paper-based Toolkit for Documenting Local Geo-spatial Knowledge.” In Proceedings of the First ACM Symposium on Computing for Development, 11:1–11:10. ACM DEV ’10. New York, NY, USA: ACM

Wooten, Joel, and Karl Ulrich. 2013. “The Impact of Visibility in Innovation Tournaments: Evidence from Field Experiments”. Rochester, NY